GREAT MEN
OF SCIENCE

GREAT MEN
OF SCIENCE

A HISTORY OF
SCIENTIFIC PROGRESS

BY

PHILIPP LENARD

formerly Professor of Physics and Director of the
Radiological Institute in the University of Heidelberg
Nobel Laureate
Rumford Medallist of the Royal Society of London
Franklin Medallist of the Franklin Institute of Philadelphia

TRANSLATED FROM
THE SECOND GERMAN EDITION
BY
DR. H. STAFFORD HATFIELD

WITH A PREFACE BY
E. N. DA C. ANDRADE
Quain Professor of Physics
in the University of London

Essay Index Reprint Series

BOOKS FOR LIBRARIES PRESS
FREEPORT, NEW YORK

First Published 1933
Reprinted 1970

STANDARD BOOK NUMBER:

8369-1614-X

LIBRARY OF CONGRESS CATALOG CARD NUMBER:

74-105026

PRINTED IN THE UNITED STATES OF AMERICA

INTRODUCTION

I COUNT it an honour to be allowed to write a few words of
introduction for the English edition of Professor Lenard's
historical studies of the great men of science. It is now
over twenty years since I worked as a research student in
his laboratory, and time has dulled many memories, but the
recollection of his inspiriting and wholehearted devotion to
the service of science, of his generous enthusiasm for the
work of men of genius, living and dead, and of his wonderful
experimental skill and resource, is still bright. Ramsauer,
Hausser and Kossel, whose names have since become
famous, were among his research students at that time, and
the physics colloquium, with Professor Lenard's illuminating and significant interjections, comments and questions,
made the pursuit of scientific truth seem an exciting and
supremely desirable quest. It was on such occasions that
Professor Lenard's interest in the history of science came
particularly to our notice. Who had first shown the way
here, what had he actually done, how was he led to do it?
– such questions, to which our professor too often had to
supply the answer himself, brought before us the greatness of past workers and the significance of their achievements. Galileo, Newton, Faraday, Hertz (who was Lenard's
teacher): such men became living figures for us, and their
tasks and successes appeared as part of an organic structure,
and not as an empty record of past times.

It is not often that one who, like Professor Lenard, has
won for himself an assured place in the history of science,
undertakes a systematic appreciation of the work of his predecessors, of the kind which we have before us in this book.

The task is one worthy of the powers of a great investigator, for he who has himself made bold advances into the unknown can best appreciate the difficulty of breaking away from accepted ideas, the labour required to bring precision into a field where nothing but vague general ideas prevail, the independence of judgment necessary in making a decision as to which facts are of fundamental bearing, which trivial. Only too often the student who has just taken his degree finds nothing remarkable in the enunciation of the laws of motion or the discovery of electromagnetic induction, ideas which are made so familiar to him that they seem obvious, and the true significance of which escapes him. He reads of the apparently simple experiments of a Rutherford or a Lenard, and feels that he could have made them himself, knowing nothing of the opinions prevailing and the apparatus available at the time, and knowing, indeed, very little about real effort of any kind. On the other hand the great investigator knows from experience the difficulty of formulating the problem in a precise form: the hold which old and established methods of thought retain: the risk of sailing in uncharted seas. It is from him that we can expect a just appreciation of the outstanding individual achievements of science, and a sure detection and appraisal of true originality of outlook. A Clerk Maxwell can best judge the greatness of a Cavendish.

In these studies of great men by Professor Lenard the reader will find a vivid sympathy, a generous enthusiasm and an illuminating criticism which brings out in lively fashion both the personality and the secret of the scientific greatness of the subject – let the life of Faraday serve as an example. The great men stand as living figures, belonging to their times, yet in many ways strangely sympathetic to ours. The author has brilliantly succeeded in presenting his subjects as individual men, of widely different dispositions, beset with the most diverse circumstances, and yet showing how all these individuals are made kin by something shared in common, the selfless search after truth. Short as the lives are, in each case the essence of the achievement is brought before us in a few words.

A strong individuality like that of the writer of this book is bound to show strongly individual judgments, and yet I venture to assert that no one will be found to deny that every name appearing in the book represents a major contribution to science and is distinguished by the originality which the author has made the password for entrance. If the character of the writer pervade the book it is rather by the circumstance that his admiration for uprightness of character and devotion to high ideals, his love of scientific truth appear on every page, than by any personal prejudices. I am glad to think that it was in England that Professor Lenard's scientific work was first recognised, perhaps, at its true worth, and I feel sure that all English-speaking men of science will rejoice to greet, in the year of his seventieth birthday, this work of the brilliant pioneer of phosphorescence, cathode rays and the ionisation potential.

E. N. DA C. ANDRADE

PHYSICS LABORATORY
UNIVERSITY COLLEGE

AUTHOR'S PREFACE

FOR a long time it had been my desire to extend my earlier reading of the lives and works of scientific investigators, and undertake more comprehensive historical studies; but I had first to pursue to my own satisfaction my aim of trying to be an investigator myself. The time for historical reading gradually arrived and the more I delved into the works and lives of great men, the more it appeared to me that much remained to be done beyond what previous histories of scientific research had given us. What most struck me in recent writings on this subject was a want of that understanding of the great men of science which, so it seemed to me, should come from a study of their life history and their behaviour. I found that these scientists – or at least not a few of them, and those the most successful – were much more above the common run of humanity than the most widely read biographies suggested. My joy was great to find that these personalities so well matched the greatness of their achievements, that they were fit to serve as examples to future generations both from the point of view of their work and from that of their lives.

I therefore thought it well to communicate this satisfaction to others as soon as my studies had progressed sufficiently for me to be sure of my facts, and also to form the whole of them into a picture of the development of science, in which individual workers would occupy a position given by the actual development of themselves and their work. For this purpose a regular plan of study was necessary, which had to comprehend not only the existing historical works on science, but also, and even to a much more important

ix

extent, the works of the men themselves. Indeed, it was above all necessary to find out which investigators really belonged to the series of great men, by whose activities our knowledge of nature has principally been developed and increased.

This study of the works of great investigators taught me that they had frequently achieved much more than they were usually given credit for. The richer the contents of their work, the more of it appeared to have been forgotten in course of time by the writers of histories and text books; or rather, the credit tended to be given to others who had later turned their attention to the subject and enlarged upon it when it was no longer new. But there can be no doubt who is the originator, when a piece of knowledge is found, quite unexpectedly, already accurately stated, and is not to be found closely anticipated in the work of predecessors. In contrast to the work of investigators to whom the above remarks apply, there are others whose works disappoint us; they prove to contain less than what is usually ascribed to them. We find, when we are acquainted with their predecessors, that their debt to these is very great; in other cases they have done no more than make assertions without having devoted serious attention to the subject; and these assertions have not been forgotten for the simple reason that others, by expending the necessary thought and trouble, have shown that something similar is actually true. But without the labours of these latter workers, the suppositions of the former would have remained nothing more than suppositions, and not have become a real addition to our knowledge. It is astonishing how certain estimates in the history of science, estimates which will not bear investigation for a moment, have continued to be held, and have proved impossible to correct. I have tried, by reading the original literature, to avoid these false estimates entirely.

If we follow up in this way the sources of our knowledge,

we are led finally to a much smaller number of names than
are usually mentioned in historical works, even those of the
smaller kind, for it appears that the writers of such books
are more anxious to name the maximum number of persons,
than to be at the trouble of forming a judgment by going
back to the sources. But since in the present work the object
was to deal with the great investigators only, this consultation
of original sources was the most important matter; furthermore, the object was to select, from our whole knowledge of
nature, those matters which are essential to a world-view
based on knowledge of nature, or which have played an
important part in the development of this world-view. This
involved the omission of a vast amount of matter, of course
equally valuable as knowledge, but not bearing upon the
particular object in view. For the only person deserving
the name of a great scientist is he who has brought forward
something entirely new, having an essential bearing on our
knowledge of nature, on our view of the universe and on the
position of man in nature. The present book gives us the
opportunity to test whether, in spite of the limitation in the
number of workers dealt with, anyone is omitted, to whom
any single piece of knowledge of first rate importance can be
traced. For in the case of each investigator dealt with, all
new knowledge which we owe to him has been mentioned, or
at least referred to; and in spite of the small number of names,
we find that the whole of our knowledge of nature, as far as
it is of fundamental philosophical importance, is discussed.
The result is that we have at once a history of the development of science, and a biographical account of the investigators to whom scientific knowledge is due.

The names of the most important and greatest investigators, around whom our whole account is grouped, will be
found in the table of contents, arranged in order of time, and
with regard to the grouping of their main achievements;
they are chosen according to the originality and general

importance of their work, according to the degree of inward
and outward difficulty which, in view of the circumstances
of their times, they had to overcome, and finally also, according to the signs of intellectual greatness which we can
recognise from the general character of their personalities.
Important work of a few other investigators, or of persons in
some way concerned in an investigation, is dealt with besides,
and a full index of names will be found at the end of the book.

Particular mention must be made of the limitation of the
matter as regards its extension to the present day. It is
obvious that living persons had to be omitted; but even
outside this limitation, a certain historical perspective was
necessary, if success was to be attained in presenting a series
of investigators judged as uniformly as possible. I decided
to take the War as the limit of time, so that no scientist
would be dealt with who had survived it. This condition
also turned out to form a possible basis, without unduly
limiting the range of the material treated, and excluding
entirely recent developments of our knowledge. This
recent development, quite in accordance with its origin,
could be touched upon in connection with the work of the
latest investigators considered, to such an extent that no
very important part of our whole picture of nature remains
without reference. Two exceptions had, however, to be
made as regards the limitation in time: Van der Waals and
Crookes, whose deaths, on account of the unusually great
age to which they lived, came after the end of the war.
It was impossible to omit them, since they were contemporaries of other investigators whose work fell within the
chosen limit of time.

The great investigators, whose works I had always
admired, appeared to me, ever since my acquaintance with
their lives, as both deserving of reverence, and supreme in
their greatness; but none the less, at first dimly and later
more clearly, I came to feel a kinship with them, and this

book now seems like a small memorial to them, which they would not despise. It is also time for such a memorial. For the number of those who are able to think in the manner of these investigators appears to be very much on the decline at the present time; and yet no other manner of thought has proved capable of more than mere momentary success, or the production of astonishment, or in the best case, as the conclusion of what has already been properly founded. Indeed, the point of view of these investigators is alone capable of leading to the solution of the questions of the moral sciences; for these questions also relate to natural processes, and nature is a unity.

This attitude of mind must therefore not be forgotten; it must again be cultivated and made to live, and no better means to this end can be found than a care for the memory of these great scientists. It cannot be hidden from anyone who looks well below the surface, that none of the achievements of the human mind are more firmly founded, none are of more permanent value, than the achievements of these men. The last fifteen years have led all those who have devoted serious thought to what has been happening, to realise that much that was earlier accepted and regarded as assured, is doubtful, nay erroneous, when seen in the light of to-day. Our view of the world has changed, and new knowledge has come to us. But these men and their contributions to our world-view have remained what they were already; indeed, the reputation and importance of them must even appear increased, inasmuch as we now see very clearly how very much further removed they were from error – and the greatest pioneers among them most of all – than their contemporaries.

The new knowledge has also put the kinship of the great investigators, which I am concerned to demonstrate in the present book from the manner of their thought, their striving, and their work, and also their whole spiritual nature, in a new light, namely as a physical relationship. There is now less than ever any doubt that a Galileo, Galvani and a Volta, a Newton, Black, Cavendish, Dalton, Davy, Faraday, and Darwin, a Fresnel and Carnot, are related to a Kepler, a Scheele, a Robert Mayer, and a Bunsen, and also to a Pythagoras and Archimedes, in the manner stated in the present book.

Let anyone who has a sense for portraits examine the plates of this volume; they will tell him the same story as is told in the text from the life and work of the men, only in an even more subtle manner than I have been able to put it. If we take the portraits of the most wide ranging creators of new knowledge – which are to be found among those we have named – I think we shall find from their examination a spiritual profit, an introduction to a world of clearer and more natural thinking, which is far from the life of to-day and from what to-day is often acclaimed as science; a world from which these investigators unselfishly brought forth, as the text tells us, all that forms the foundation of our present progress in knowledge and technical power. At the same time there can be little doubt that these men of science would find little satisfaction in the achievements of our so-called civilisation, so far as they do not really make our life better, but that they would rather seek for progress in morals and true culture, which might have come about as the result of their manner of thought and their scientific achievements. Such advances might yet develop, if these men should come to be understood by us as regards their way of thinking and working, and exert their rightful influence, to a greater extent than has hitherto been the case. I hope that I shall have been successful in contributing towards this end.

The publisher, Mr. T. F. Lehmann, Munich, is to be thanked for much help; he has given me every assistance, particularly in the production of the portraits.

The Deutsche Museum in Munich has provided us with Figures Nos. 4, 25, 27, 29, 37, 41, 57, 62.

From the Corpus Imaginum of the Photographische Gesellschaft I have taken the following figures: – 6, 11, 16, 19, 26, 30, 31, 33, 39, 40, 47, 50, 52, 53, 55, 61.

Figure 32, Davy (page 194), is taken from an engraving, after a painting by Sir Thomas Lawrence, found in Davy's *Reminiscences*, published by his brother. Figure 46, Tyndall (page 282) is from a drawing by George Richmond. Figure 43, Faraday, with the 'heavy glass' (page 262) is from the book *Life and Letters of Michael Faraday* by Bence Jones. Figure 44, Weber (page 263) comes from Zöllner's memorial of Weber. Figure 45, Mayer (page 282), is from Weyrauch's Centenary volume to the memory of this scientist.

My thanks are also due to friends and colleagues, who kindly assisted me in collecting data and portraits, and also in advance to those who will help me further as I hope in the future to improve the work.

P. L.

HEIDELBERG

CONTENTS

LIST OF ILLUSTRATIONS

xix

GREAT MEN
OF SCIENCE

GREAT MEN OF SCIENCE
AND THEIR WORK

PYTHAGORAS OF SAMOS
About 570–496 B.C.

PYTHAGORAS, whose geometrical theorem is known to every-one, and who was already acquainted with the harmony between notes produced by strings having simple relations of length to one another, is the earliest investigator whose name is associated with definite advances in knowledge. He was not only the first to make the advances we have mentioned, but also, at that early date, the first to state and emphasise the essential importance of measurement and number for all knowledge of nature. He was the head of a great school devoted to the restoration of the moral purity of past times. This school was already familiar with the idea of the spherical form and daily rotation of the earth. From this we can judge how far those times were advanced in unprejudiced thought concerning nature; two thousand years later Galileo championed this idea of the rotation of the earth, not against justifiable doubt concerning the scientific correctness of this conception, but against the overwhelming tyranny of superficial and obviously inferior minds. Never-theless the time of Pythagoras must have been the beginning of a decline from a previous culture, which is shrouded in the mists of time; for his school aimed at the salvation of *past* greatness. This decline appears not only in the violent end of Pythagoras and his school, and later in the disgraceful end of Socrates and in Plato's vain striving, but above all, in the total destruction of all cultural achievement in the thousand years which followed.

Before this period of desolation commenced, three great figures appeared: Euclid, Archimedes, and Hipparchus.

EUCLID

330 (?)–280 B.C.

EUCLID appears as the first comprehensive investigator of space, the three-dimensional space with all its properties, in which we live, and of the forms and structures which are possible in it. He founded all the essentials of geometry. But he is at the same time also the founder of elementary mathematics. He had already demonstrated the propagation of light in straight lines, and discovered the law of the formation of images in mirrors. Much of this, it is true, can only be surmised from relics of his works of doubtful authenticity; very little is known to us regarding his life, and no portrait of him exists. Little more can therefore be said here.

ARCHIMEDES

287–212 B.C.

ALMOST the same is true of Archimedes, but his writings have been better preserved. They tell us that he had already formed a conception of force for cases in which it does not produce motion, that is to say, where there is equilibrium between several forces (statics). He laid down the conditions of this equilibrium when the forces act upon rigid bodies, and thus developed the important subject of machines. The lever and set of pulleys became available for the multiplication of power to almost any desired extent, and in a predictable manner, and striking instances of the successful applications of these principles, as, for example, to the raising of heavy ships, are related. The Archimedean screw is still named after him to-day.

He also developed our knowledge of the centre of gravity, which allows us to understand and predict how the weight of a heavy body acts when it is supported in a given manner. He is also the founder of the theory of the equilibrium of liquid bodies (hydrostatics); the Archimedean principle of the upward force on bodies immersed in liquids, which also determines the equilibrium of floating bodies, constantly finds application. His work 'On Floating Bodies' presents us with a mass of finely thought out details; the idea of specific gravity, or density, is already fully worked out.

He also greatly advanced geometry and mathematics. He discovered how to calculate the circumference of a circle, by means of inscribing and circumscribing polygons, starting with the hexagon, and repeatedly doubling the number of sides. This method is one which gives the number later designated by the Greek letter π to any required degree of accuracy. He also developed the theory of conic sections, which later became so important for astronomy, and likewise the calculation of the surface of ellipses and parabolas, and the calculation of the volume of a sphere and of other solids. He is also known by his investigation of the Archimedean spiral named after him. His profundity is shown in the fact that he was the first to actually grasp the concept of infinity, as opposed merely to the very great, a fact clearly shown by his essay known as the 'Sand-reckoner.'

HIPPARCHUS OF NICEA
160–125 B.C.

HIPPARCHUS was the first of the great observers of the heavens who made actual measurements. Of him also no portrait exists. He followed the movements of the sun, moon, and planets by angular measurement, and deduced from his results much improved values for the length of the year, the inclination of the ecliptic and the eccentricity of the moon's orbit. He discovered the very slight inequality of the length of the apparent solar day, and the precession of the equinoxes. He further recognised that the distances of the sun and the moon from the earth are variable. He attempted to determine these distances, expressed in semi-diameters of the earth, and succeeded very well in the case of the moon. In the case of the sun, however, he was not able to attain the accuracy necessary for such great distances, and his result was much too small, but he found correctly that it was very much greater than the distance of the moon. His star catalogue gives the position of over a thousand fixed stars. He was thus the actual founder of measurement in astronomy, and was only surpassed much later by Tycho. He thus provided the foundation upon which Copernicus built. His life is unknown to us; even his achievements are almost all known to us only through the later work of Ptolemy.

These great men were followed by a period of more than fifteen hundred years, which was practically barren of all results for scientific investigation, and indeed, for the advance of knowledge generally (about 150 B.C. to 1500 A.D.). In this period no single name of equal importance can be mentioned. Men had forgotten how to think of space and time,[1] and the

[1] What comes to us from this period, mainly attributable to the Arabs, does not concern us here, since we are only considering important and essential progress in science. Art also passed through a completely barren period; but architecture and poetry attained to high development much sooner than science.

processes taking place in them. It is quite clear that the men necessary for this were missing; they were lost, they were no longer born. Why was this ? We now regard it as proved that the original habitation of races fitted for such achievement is the north of Europe, where they had been developed through many generations in the hard school of the Glacial Period. It was from there that the ancient Greeks had come who then, in the South, found leisure and ease for a development which reached its height in Pythagoras, Euclid, Archimedes and their contemporaries of equal ability in art. But their posterity began to degenerate. As far as all our present knowledge can tell us, their race did not leave descendants, but was hopelessly ruined by the influx of Asiatic and African elements.[1] We have seen that Pythagoras was already forced to strive for the restoration of the purity of past times. It was necessary that new men from the North, the only home on earth of the seekers and bringers of light, should again come southwards. Further, this race was gradually able to develop in its original home, assisted perhaps by an improvement in the Northern climate and progress in the invention of means to combat its asperities.

The thousand-year period of degeneration, as regards the gradual rise of a new culture, was distinguished in science and intellectual activity generally by its unproductive concentration on two works received by it from antiquity: Aristotle and the Bible. Both of these were given totally false values, and misunderstood.[2]

[1] There can be no doubt that physical racial degeneration was preceded by moral infection, and also that an excessive number of wars produced their effect; but these facts do not alter what has been said above.

[2] As regards the Bible, this is true for the general public to-day. All that has disappeared from the general mind is the contradiction of natural fact arising from literal interpretation; historical errors and contradictions of vital law still continue to produce their effect. For in these writings, parts of the highest value are thrown together with utterly valueless matter of different origin, and people are still afraid to make this fact generally known. The resulting misdirection of intellectual

Furthermore, they were effectively protected from all criticism by physical force, which was all the easier since these writings were not readily accessible; and their influence was therefore extremely injurious. Everything that then happened, even when it already showed strong signs of new progress, was in this way guided amiss, and thus robbed of its effect on intellectual progress. We may remember, for example, the century-long misdirection of the heroic spirit through the Crusades.

Aristotle, who lived before Euclid (384–322 B.C.), appears to us, when we compare him with the great men who preceded and succeeded him, mainly in the character of a prolific author; he is an obvious symptom of degeneration. He assumes the rôle of omniscience in all matters, and claims to present the last word of wisdom. This at least was the impression he made on his feebly gifted successors, and hence his works were saved to a surprising extent from destruction, although they were hidden for a time. We know no more of his personal life than of that of Euclid and Archimedes.[1] The fact that he was followed almost blindly for a thousand years, did not depend upon him, but upon the men who lived during that period. All his statements concerning natural phenomena could easily have been tested; in fact very easily, since they mostly related to facts of our everyday surroundings, which anyone with leisure can examine with any degree of accuracy. But people had

attitude affected, at a time when the actual facts were completely unknown, even men of science of the highest rank such as Kepler and Newton, who were faithful seekers in the world of spirit. Thus Kepler for example, in the introduction to the fifth book of the *Harmonies*, innocently remarks in his joy over a great success: 'Yes ! I have stolen the golden vessels of the Egyptians.' Or Newton, who attempts in his latest writings to arrive at the exact construction of the temple at Jerusalem as a matter that men of his character must necessarily regard as precious and important to salvation.

[1] It was reported that Aristotle, who came from the south of Greece, was small and slim in figure, dainty in his behaviour, and inclined to sarcasm in conversation.

obviously become incapable of such activity. Besides, for a long time it was forbidden by force to cast doubt either upon Aristotle or upon the Bible. The University and the Inquisition took care of this, which fact is a further proof of the low level to which intellectual activity had sunk.

The healing effect of the passage of centuries, which allowed men of sufficient purity of blood again to reach dominance[1] was first seen in two events of a more superficial character, though art in the form of the Gothic cathedrals had already developed. Gutenberg invented printing in the year 1440; Columbus in 1492 began his bold and protracted journey to the West, and brought back the astonishing news of the circumnavigability of the earth, which showed it to be a sphere freely suspended in space; with living beings upon it with their feet towards us (Antipodes). Hitherto it had been forbidden even to speak openly of the possibility of their existence; while Gutenberg's achievement now made it much easier than before to bring living and enquiring spirits into contact with one another, Columbus' discovery showed the assumption of a fixed and immovable earth, one of the chief doctrines of Aristotle and the Bible, to be at least extremely doubtful.

Upon this followed, step by step, a new era of investigation, the main steps of which we shall now follow in connection with their chief representatives.

Before doing so, I should like to express my opposition to a point of view which is frequently put forward to-day; the view that this new era of investigation depended upon a new method of investigation, namely observation and experiment; and that Tycho, Stevin, and Galileo were fundamentally different in type from Archimedes and Hipparchus.

---

[1] This effect was certainly aided by the spread of a wider knowledge of Greek literature (after 1453); but these writings were also previously not entirely lost to knowledge, and the fundamental reason for their temporary disappearance still remains the absence of persons capable of properly understanding them.

This is not the case; they merely resumed and carried on a method which had been begun, but then allowed to go to waste, or merely to revolve around uncreative changes. Even Aristotle, mentally distorted as he was, had made observations. When for example, he states that light bodies fall more slowly than heavy ones, he had without doubt observed the fall of feathers and stones. Stevin and Galileo certainly went about their work in a different manner from Aristotle, since they again took seriously both the multiplicity and also the quantitative aspect of observation; but Pythagoras had already expressly pointed this out, and Archimedes and Hipparchus had acted in the same manner. Stevin, Galileo, and the great men who followed them, were certainly very different from Aristotle with his hasty generalisations and easy certainty; but in this respect also they were only like their true forerunners. They passed only gradually from observation of what offered itself directly to them, to an increasing degree of true experiment; that is to say, to carrying out observations under carefully thought-out and favourable conditions, and with the assistance of specially constructed appliances. In this way, Galileo arrived at his experiment with the fall of bodies on the inclined plane and his observations of the heavens by means of the telescope; but Archimedes had also experimented when he immersed various bodies in water.

The progress of the new epoch as compared with the old therefore consists only in the gradually increasing and intentional multiplication and refinement of observation, and in the increasing use made of conditions intentionally arranged for the production of the phenomena to be observed; as for example, when Galileo investigated the movements of pendulums swinging with different arcs, or constructed of different lengths. This was not essentially different from the procedure of Archimedes when he varied the length of the arms of the lever, or imagined them as varied, with reference to his previous observations. Progress in this direction came about of necessity, as soon as the possibilities were exhausted of drawing conclusions from casual and remembered observations of processes taking place by themselves all around us. The essential advance was in the direction of a more consciously and advantageously chosen alteration of the conditions of experiment; a step which proved fertile in results. Progress in this manner of investigation had already commenced in antiquity; it was interrupted, together with all other investigation, for fifteen hundred years, to be again taken up more particularly by Galileo – but even in his case not suddenly, but only step by step – and again developed further. The assertion that the investigation of nature in the form of experimental science did not exist before our era, is contradicted by the example of Hipparchus alone, who may be regarded as the actual founder of astronomy based upon comprehensive observation and the collection of experimental data.

LEONARDO DA VINCI
1452–1519

LEONARDO was the first great man of the new epoch. He was born in Vinci near Empoli, not far from Florence, and there received his early training as a painter and musician. In the latter capacity he was received at the court of Milan, where he painted his famous 'Last Supper.' He was no less active as an artist, and as adviser during the construction of the fortifications and canals. He also appears to have founded an academy for art and science in Milan, of which he himself formed the centre. War put an end to all this splendid development after nineteen years. Leonardo then returned to Florence. He later sought once again a field for his

activity in Milan, and then in Rome. Three years before
his death he migrated to France at the invitation of the King,
and died there at the age of 67.

We see that to know him only as a painter, in which capacity he is generally known, is not to do him justice. He was
in a quite special degree an investigator of nature, as appears
from the very extensive manuscripts which he left behind.
These are legible with difficulty, for he generally wrote with
his left hand. Illustrated with many drawings, they offer a
mass of natural knowledge derived from his own observation,
which points far into the future. He published nothing of
this in print himself; but he must have expounded most of
it in his lectures in the academy which he founded. We
find there for the first time ideas concerning the details of
the motion of falling bodies. As the result of experiments
with pieces of wood and lead falling from a tower, he recognised that fall was an accelerated motion, although he
discovered nothing else of importance in this direction. The
dim disc of the moon which accompanies the bright sickle
shortly before and after new moon, was explained by him
quite correctly, one hundred years before Galileo, as due to
the reflected light of the sun-lit earth. He examined the
lever and other machines more closely than Archimedes, and
in considering forces acting at an angle upon a lever, he
already formed a correct conception of turning moment. He
also distinguished between sliding and rolling friction, regarded friction quite rightly as a special form of force, and
determined its independence on surface area. He was probably the first to clearly distinguish the concept of work from
that of force, inasmuch as he remarks that work implies a
distance moved in the direction of the force. He observed
waves in water, regarded sound correctly as wave motion in
the air, and concluded from observations of echoes that
it has a definite velocity of propagation. He noticed the
rise of liquid in narrow tubes, and regarded it as a special

phenomenon. He further convinced himself that air consists of two components, one only of which supports combustion; also that when this constituent is used up, no animal can survive in the air. He wished to use the lower density of warm air to cause balloons, made airtight with wax, to rise.[1] He constructed or planned hygrometers, grinding machines for concave mirrors, flying machines, parachutes, diving dresses, and many other things.

Newton in his old age once compared himself to a boy who, playing on the seashore, found a few more beautiful pebbles than his companions; to which we must add that he also took care of his finds and worked upon them with great success. Leonardo found and collected pebbles in great numbers and of many different kinds. The very special ones, which he worked upon with great intensity, are his immortal paintings, and not his discoveries as a scientist. But he was obviously of the same type as Newton; he took hold of what nature offered from pure joy of creation and knowledge, and faithfully did his best with it as far as the state of the times and the unfavourable nature of surrounding circumstances allowed him. The fact that Leonardo could never satisfy himself as regards the execution of his work is shown not only by the small number of his paintings, all of which are masterpieces, but also by the fact of his scientific work being left in manuscript only. Though he regarded the latter as unfinished, he was still very far ahead of his times in regard to it.

It is reported that Leonardo was distinguished by his beauty, strength, and bodily agility, and by an inexplicable attractiveness in everything that he did; also that with all

[1] Only three hundred years later, when the epoch of the steam engine had already begun, was this idea carried out on a sufficient scale; the brothers Montgolfier, French paper manufacturers, sent up some animals in a balloon ten metres in diameter, in the year 1782. These returned safely. In the following year the first ascent by human beings was made in a similar but much larger balloon. Hydrogen, which had not long been discovered, was first used at about the same time.

his reserve, he showed great brilliance and wit in his conversation. He always had a great affection for animals.[1]

NICOLAUS COPERNICUS

1473–1543

COPERNICUS was born at Thorn on the Vistula, and studied medicine, mathematics, and astronomy at Cracow. His name was usually written by himself as 'Coppernic'; but in his work De Revolutionibus we find 'Kopernikus.' He lost his parents at an early age and was sent by his maternal uncle, later Bishop of Ermland, to monastic life. He had to travel a great deal, which took him to Vienna and Rome, and he continued his studies very extensively at various Italian Universities. At about the age of 27, he gave public lectures on astronomy in Rome, but then studied canonical law and finally became canon of Frauenburg near his home. This gave him the possibility of devoting his life to undisturbed study for many years, particularly at first, and then again as an old man. But he also at times played the practical part of bailiff to the extensive lands of the cathedral, at the Castle of Allenstein, and performed this function during difficult periods of war time with care and firmness. He was for a considerable time the representative of the Cathedral Chapter at the Prussian Landtag, acted as an expert when a new

[1] Leonardo's life, art, and personality, have been described in The Mind of Leonardo da Vinci, by E. MacCurdy (1928), who has also edited his note books, and published an abridged edition of them. The reader may also be referred to Merezhkovsky's famous novel The Forerunner, which is a study of Leonardo's personality. In the German language, we have the work of Woldemar von Seidlitz (2 vols. Berlin, 1909). A detailed appreciation of Leonardo as a scientist who built up his ideas very decidedly upon experience and observation, is to be found in E. von Lippmann's Abhandlungen und Vorträge, vol. I, pp. 346–375 (Leipzig, 1906). See also Leonardo als Techniker und Erfinder, 2nd ed., by F. M. Feldhaus, 1922.

mint was founded, and also had a high reputation in medicine.

Copernicus was a contemporary of Luther. He did not join the reformers, but always took a tolerant attitude towards them, even when in danger of coming thereby into conflict with his ecclesiastical superiors. Copernicus spent thirty-six years upon his famous work *De Revolutionibus orbium coelestium*, up till nearly the time of his death. He started from the wish to calculate the motion over the heavens of the sun, moon and planets, in a simple and trustworthy manner. As foundation he took the traditional knowledge and views which at that time were to be found in the thirteen books of the *Almagest*. The compiler of this, Ptolemy (70–147 A.D.), took his main data from Hipparchus, whose observations were of high quality, and in so far Copernicus was well equipped. But in the *Almagest* the earth is assumed to be at rest, which at that time also was regarded as beyond a doubt. But Copernicus could not succeed, on the basis of this assumption, in calculating satisfactorily the motion of the heavenly bodies as actually observed from the time of Hipparchus up to his own time; the many circles and epicycles which had to be assumed as the path of all bodies moving around the earth became extremely complicated, and the complication became greater and greater as Copernicus attempted to bring the actual observations into his calculations. For example, the orbits had to be assumed as arranged eccentrically about the earth, which resulted in the abolition of the fundamental idea, that of the central position of the earth. On the other hand, everything became quite simple and comprehensible as soon as Copernicus tried abandoning this fundamental idea completely, allowing the earth to revolve around the sun like the planets, and also to turn once daily upon an axis which remained always parallel to itself. From the point of observation of an inhabitant of the earth regarded as moving in this manner, the actually observed positions and motions of the other planets resulted in a simple manner; they were merely assumed to be likewise moving in certain paths around the sun, but of different diameters. The fixed stars were assumed to be the background upon which the motions were observed, and to be very much farther away from us than the sun and all the planets. The only body that still remained as revolving around the earth was the moon, but this, as Hipparchus had already found, is much nearer to us than all other heavenly bodies. This view of the planetary system as a stationary sun and moving earth, was already that of the school of Pythagoras; which is a proof of the lofty flights of which those minds were capable, although in their case it was not the result of firmly founded investigation. Copernicus was able to make it so, since he had Hipparchus' thorough observations at his service, and was not satisfied until completely certain that other possibilities, suggested before his time or occurring to himself, did not agree in the same simple manner with observed reality, as did this assumption of the moving earth. But this assumption then became for him a reality recognised as such; he shows himself as a true investigator of nature when he states that he has found that, and how the earth moves: 'All this, however difficult and almost incomprehensible it may seem to many, and however much it may be opposed to the ideas of the great majority, all this we shall, with God's help, make clearer than the sun in the course of this work, at least for all those who are not completely devoid of all mathematical knowledge.'

He also expressly draws a very remarkable conclusion from his discovery. If the earth describes yearly a path of great diameter around the sun, the fixed stars cannot by any means be so near to us as was thought at that time. For there are no visible signs of a semi-annual change in distance, as for example, a variation in magnitude or brightness. The not very large sphere, to which all fixed stars were imagined as fastened, and which enclosed the Universe, could not correspond with reality; on the contrary, the fixed stars receded to a distance which, in comparison with the already great distance of the sun, must appear as immeasurably great. Mankind had suddenly approached the conception of the infinity of the universe.

A new picture of the world had come into being, and for the first time, it was supported by figures based upon experience. This picture given by Copernicus has remained; it has only been defined, as always in science, by further gradual development, and thus been linked with more profound ideas. Tycho, Kepler, Galileo, Huygens, Newton, are the names with which this further development, which lasted almost for two centuries, is associated.

Copernicus began his work early and finished it about 1530 in all essentials, but long kept it back, not for fear of admitting to knowledge, no matter how objectionable and inacceptable it might appear to contemporaries, but for fear of the noisy interference of incomprehending persons.[1]

He was in doubt 'whether I shall make my work known or whether I shall follow the example of the Pythagoreans and only impart its contents by word of mouth to my friends.' It thus came about that only on the day of his death was he able to take in his hands his first printed pages. He thus neither experienced the indifference with which this purely scientific work was at first received, nor the persecution which later on became its lot from the ecclesiastical powers.

The first powerful effect of Copernicus' work upon an allied mind, and perhaps its most profound effect of all, appeared about forty years after its publication in the case of Giordano Bruno, the Dominican of Nola, who, like

---

[1] The first and somewhat cautiously worded preface to the first printed edition of Copernicus' work was not written by him and did not have his approval, but was added by the publisher with an obvious eye to the danger ahead, but it did not prevent the work being later forbidden. In other respects there are several notable departures from the original manuscripts (which still exist) in this first printed edition.

Luther, could not remain in a monastery. At a time when no general recognition of it existed, he praised publicly, in word and writing, the discovery of Copernicus as a deed which would relieve humanity and form the starting point of a new era of thought and investigation. For him, generalisation is the way to wide views. What he thus foresaw, was incapable of demonstration in his time, and thus does not represent a progress in knowledge; but he strives to attain the last limit of knowledge, which even to-day can only be a matter of surmise, and which, though going far beyond Copernicus, must still be based upon his investigations, and hence finds its place here. He abandons the sphere of the fixed stars enclosing the solar system, which had already been greatly enlarged in size by Copernicus, dissolves it completely, and sees in the fixed stars for the first time, suns like our own, distributed in infinite numbers freely in space. All these suns he sees as surrounded by planets, by earths like our own, and like it, peopled by living beings. The space of this universe, in which not only our earth, but also our sun, no longer have any pride of place and central position, he considers, for the first time with good reason, as unlimited, infinite. Hence dwelling places for living minds are distributed everywhere in this universe, though they are separated by vast distances which cannot be physically surmounted. 'There is only one heaven,' he says, 'an immeasurable domain of light-giving and illuminated bodies'; the Godhead is not to be sought far away from us, since we have it near to us, yes in us, more than we ourselves are in us ; so must the inhabitants of other worlds not seek it in ours, since they have it in their own and in themselves.'

For this the Inquisition and the Pope sent him to death at the stake. He was burnt in the year 1600 on the Piazza dei Fiori at Rome, remaining to the end full of disgust for a Christianity which cannot be made to agree with knowledge such as that which he owed to Copernicus.

LEONARDO DA VINCI

NICOLAUS COPERNICUS

TYCHO BRAHE

1546-1601

COPERNICUS himself left behind him the wish that the position of the planets calculated by his methods should in the
future be compared with their actual positions, for the
purpose of testing the correctness of his statement regarding
the paths of the heavenly bodies. The calculation was performed by Erasmus Reinhold (1511–1553) in the *Prutenischen
Tafeln* ('Prussian Tables'), which represented a great
improvement as compared with the astronomical tables
hitherto in use, and became of importance also for the
calendar. Copernicus' work was thus proved to be correct;
but the agreement with actual observation was not perfect.
Thus Mars, for example, was occasionally found as much
as two degrees (four diameters of the full moon) distant
from its calculated position. Discrepancies of this kind might
depend upon the original data, which in Copernicus' case
were in part derived from Hipparchus, and the admirers of
the great advance made by Copernicus were inclined to
assume this to be the case; but it was also possible that the
actual path of the planets might still exhibit peculiarities
which were unknown to Copernicus.

At this point, therefore, an unprejudiced and accurate
observer might again find the means of making further
progress, and in this capacity Tycho Brahe showed the
way to magnificent new discoveries. He refused to accept
Copernicus' system without question; he was simply
anxious to determine as accurately as possible the true path
of the planets among the fixed stars; to which end, the positions of the latter had also to be determined with greater
exactness than Hipparchus had done, and Tycho proposed
to make these determinations to a degree of accuracy
hitherto unattained. This aim he followed tirelessly during
almost thirty years of work. His measurements of stellar

positions were made by means of new and very elaborate apparatus invented by himself, and reached the greatest exactitude obtainable without telescopes. For example, he took into account, and determined for the first time, errors in the division of his circles and the refraction of the air. His measurements, which are contained in extensive tables, give us the position of stars with an accuracy of half a minute of arc (one-sixtieth of a full moon diameter) while Copernicus himself had regarded observations twenty times less accurate than these as especially successful.

Tycho was the son of a Swedish nobleman, and began by studying law in Copenhagen; but then, following his own inclinations and in opposition to the wishes of his family, he turned to astronomy. At the age of sixteen, he went to the University of Leipzig, then to Wittenberg and Rostock; he also carried on studies in alchemy at the same time. It is possible that these enabled him to make the alloy of silver with which he ingeniously replaced the upper part of his nose, which he had lost in a duel. In any case, great experience in the working of metals was very valuable to him in the construction of his instruments. His main life work, to which we have already referred, was rendered possible for him by the favour of princes. King Frederick II of Denmark, on the recommendation of the Landgrave William IV of Hesse, presented him with the island of Hven in the Cattegat, and built for him, when he was thirty years of age, the observatory of Uranienburg, as well as the house Sternenburg for him, his assistants, and scientific guests. The observatory was splendidly equipped and became very famous.

He worked here with the greatest energy for twenty-one years. At the end of that time his work was disturbed. Frederick II died; four regents carried on the government during the minority of his successor. Tycho fell out with one of them, and thus made it easy for his enemies, of which so energetic and proud a man must have had a large number, to undermine his position at the court; so he decided to leave his Uranienburg, and Denmark itself. Thereafter the observatory soon fell into ruin. After two years of life in straitened circumstances, mainly in Rostock, Tycho at last succeeded in finding a new opening. He established a connection with the Emperor Rudolf and became his astronomer, astrologer, and alchemist. He was given a house in Prague, and the Schloss Benach near Prague to live in, and sufficient means to build a new observatory. But he was only to enjoy this new field of work for a single year. He died after a short illness at the age of only 55 years. His rich observational material was passed to Johannes Kepler, who was his assistant in Prague. We are told that one of his last remarks was 'Ne frustra vixisse videar' (May I not seem to have lived in vain).

Tycho was not merely the exact astronomical observer Though the actual application of his life work on the planetary motions was reserved for Kepler, he nevertheless himself drew very important conclusions. His first essay, which made him widely known, related to the new star which suddenly appeared in 1572, and was discovered by him; he decided from the want of proper motion of this star, that it belonged to the fixed stars, and emphatically stated that the world beyond the planets cannot by any means be unchangeable, as was then generally assumed. He also observed comets. He noticed that the position of a comet among the stars is the same for observers on the earth very widely separated from one another, and concluded from this that comets are very distant heavenly bodies, and not, as had been hitherto assumed, luminous formations in the earth's atmosphere.

He further found from the motions of the comets that their paths extend from very far away to very near the sun, which clearly showed that the space of the heavens allows of free motion between all these orbits of the planets. There was no trace of spherical bowls of crystal, by which the planetary motions were still regarded as being guided. Nevertheless, Tycho still found it difficult to assume that the earth moved; he saw too many objections to this idea. It is unjust to interpret this as weakness, in the case of a man whose life work gave us the means of completely meeting all possible objections to this theory. What was missing was 'dynamics,' the knowledge of the laws of motion of matter and the forces acting upon them, which apply equally to the stone when thrown upon the earth, to the whole earth, and to all the other heavenly bodies. The road to this end was a long one; and was only completely traversed by Newton.

Tycho had reddish hair and was a great friend of animals.[1]

SIMON STEVIN
1548–1620

THIS scientist, who is all too little known, was according to Leonardo the first to continue consciously Archimedes' studies in mechanics. He also quite expressly based his work upon this master of antiquity. Copernicus and Tycho attempted to discover the actual motions of the heavenly bodies which are presented to us for observation, but are entirely beyond our interference, and to understand the world on this basis; Stevin again turned to earthly machines; inclined planes, levers, pulleys, first engaged his attention,

[1] As a result of a duel, Tycho at the age of twenty lost part of his nose. The missing piece was replaced by some kind of metal contrivance made of gold and silver. As Dreyer says: 'The various portraits which we possess of Tycho show distinctly that there was something strange about the appearance of his nose, but we cannot say with certainty whether it was the tip or the bridge that was injured, though it seems to be the latter.' The astronomer J. L. E. Dreyer gave us an excellent life entitled *Tycho Brahe, a Picture of Scientific life and work in the sixteenth century* (Edinburgh, 1890).

and he found laws of their action and connection of the most general importance; and from these he proceeded to advance very much further.

He began as tax-collector in his native town of Bruges, but left this activity already at the age of 23 years, and then travelled through Germany, Poland, and Sweden, and also studied at the University of Leyden. His extraordinary power of dealing with mechanical questions resulted finally in his becoming controller of the land and water construction of Holland, and quartermaster-general of the Dutch army. Departing from the usual custom of the time, he published his scientific studies not in Latin, but in his native tongue, Dutch; and they were then translated into other languages.

His main approach to the doctrine of the equilibrium of forces (statics), which he brought to perfection in all fundamental points, was the inclined plane. The picture of the triangular inclined plane standing upon a horizontal base, with the endless chain around it, is found on the cover of one of his chief works, with the inscription 'Wonder en is gheen wonder' (A marvel and yet no marvel). The marvel in this case is the simplicity of the rule, that the ratio of reduction in force on the inclined plane is equal to the ratio of the height to the length of the plane. It is shown that this law is not a marvel; for it is proved to follow from the fact, which appears to us as self-evident, that the endless chain in question would never of itself start to slide on the plane by virtue of its own weight, however frictionless the contact might be. Stevin exhibited for the first time, in a fully worked out and important example, a mental experiment; that is to say, an experiment which does not need to be carried out in reality, since the result of it is already sufficiently in our possession, and hence already available in consciousness. The chain will not start to move. For if it did so, it would have to continue to move indefinitely, since in either direction of motion the same distribution of

gravitational forces would always exist, and the result would
be perpetual motion. Stevin, who had a practical knowledge of machines, takes for granted that perpetual motion,
an arrangement continuing to move without external aid, a
machine going by itself, is impossible; and hence likewise
the motion of the chain on his inclined plane. The chain
must therefore be in equilibrium, and we are thus immediately given the conditions for equilibrium on the inclined
plane, and the ratio of forces, if we imagine the freely hanging part of the chain as abolished. Stevin thus connects
together the result of our experience that a perpetual motion is impossible, with the ratio of forces on the inclined
plane, and then further with the parallelogram of forces,
which first appears in his work and is there applied in very
many ways. This firm linkage of different laws, which can
be separately tested by experience, in such a way that they
all stand or fall together, is highly characteristic of exact
scientific research – the peculiar certainty of its results depends upon such interconnection – and Stevin affords us the
fundamental example which we have just described.

He connects this knowledge with the laws of the other
machines, such as sets of pulleys and levers, and thus arrives
at a connection with the work of Archimedes and Leonardo,
and in fact, at a complete understanding of all kinds of
machines, that is, of all arrangements, which allow the magnitude, direction, and the point of application of given forces
to be changed as required. The study of pulleys leads him
for the first time to a real understanding of the 'principle of
virtual displacement,' which says, that in the case of small
displacements in the neighbourhood of the position of equilibrium, force and distance are inversely proportional to one
another, or that what is gained in reduction of force is lost
by increase in distance. This principle is true of all
machines, even when they work with liquid, as in the subsequently discovered hydraulic press. It was of the greatest

importance in making clear to us the concept of work. We
see that the amount of work, as measured by the product of
force into distance, remains unchanged in the case of all
machines.

Also in the case of the equilibrium of liquids (hydrostatics),
we are acquainted with investigations by Stevin of fundamental importance, and dealing with practically all important points. He deduces the laws of distribution of pressure
in liquid, and the pressure at the bottom of the vessels,
whereby he again starts from the idea of the impossibility
of perpetual motion, and adds a further rule, according to
which the equilibrium in a liquid is not disturbed, when
a part of it becomes solid (without change of volume). He
thus arrives at the important fact, which appears a paradox,
that the pressure at the bottom of a liquid is independent
of the section and shape of the liquid column producing it,
the height and specific gravity of the liquid being the sole
factors determining the pressure. He tested and proved this
law by means of a pair of scales, one pan being the bottom
of a vessel of liquid, to which he could give different shapes.
In his consideration of liquids in communicating tubes, and
of floating bodies, he again has points of contact with Archimedes and Leonardo, but he goes beyond the Archimedean
principle, inasmuch as he does not merely determine the
magnitude and position, but also and for the first time, the
point of application of the buoyancy, the 'metacentre,' which
remains fixed, even when the floating body is moved within
certain limits. He made use of this new piece of knowledge
in the construction of ships.

Stevin also introduced calculations by means of decimals,
but at the time these did not find further adherents. The
fact that he studied not only Archimedes, but as was usual
at the time, Aristotle, is shown by his experiments on the
fall of bodies of different weight, which, in opposition to
Aristotle, he found to fall at equal speed. He says in a work

which appeared in 1605: 'Take, as Professor Jan Cornets de Groot and I have done, two lead balls, one ten times larger and heavier than the other, and allow them to fall from a height of thirty feet on to a plate or other object, upon which they strike with sufficient noise, and it will then be found that they strike the plate at the same moment, so that both sounds seem like one.' With this he goes considerably beyond questions of equilibrium (statics) and engages upon an investigation of the process of motion (dynamics), wherein Galileo then followed him with decisive results. What here distinguishes Stevin and Galileo from Aristotle in a very fundamental way, is their obvious inborn and lively feeling for the supreme importance of simple processes of motion such as the mere fall of bodies, these being roads to the understanding of all phenomena of motion. Aristotle must have been quite devoid of this feeling. For him the motion of falling was of no particular interest; he dealt with it as with the mass of other matters which he undertakes, without having properly examined any one of them, although Pythagoras already existed as an example of how investigation begins with simple things, which nevertheless must be treated with quantitative exactness.

GALILEO GALILEI
1564–1642

HERE we come to a man of the highest genius, who was able to exert in many directions such an influence on science as to change the whole position. Above all, he was the founder of the doctrine of the motion of matter (dynamics), which he turned from scattered small beginnings and suppositions into a finished science. Since all that happens to tangible and weighable bodies as far as they are without life, from the smallest to the greatest, from the atoms to the planets and suns in cosmic space, is only motion of them or their parts, Galileo thereby founded the whole physics of matter. The investigation of cases of equilibrium (statics) in which motion does not occur had no doubt to come first, since it is simpler, and we saw how this had been studied from the earliest times down to Stevin. But the general and most important case is one in which the conditions of equilibrium are not fulfilled, and we have motion taking place.

All the writings of Galileo show us that before his time no one had been able to recognise in detail the character of even so simple a motion as that of a falling or projected stone, and still less to calculate the course of such motion in advance; it had always remained a dimly realised transition, from one position of equilibrium to another, the expression of an imagined striving of every body to reach its 'natural position' where it belongs. How great is the difference between this stage of knowledge and our present ability to predict all phenomena of motion, as soon as the masses moved and the forces acting are sufficiently known! This forms the whole foundation of present-day mechanics, acoustics, and theory of heat, which together make up the physics of matter. As a later and further part of physics we have the physics of the ether, which goes beyond Galileo; but even the knowledge that this is the case has only come to us quite recently, so that until a short time ago, Galileo could be regarded as the founder of the whole of physics. His achievements in detail may be gathered from the following account of his experiences; did this refer to several men and not to one alone, they would all be individually among the greatest scientists.

Galileo came from an old and respected Florentine family. His father appears to have been a merchant, but was also a teacher of music, and known as author of works on music. His son also inherited a love of this art, and even unusual skill with the lute. He also showed much talent for drawing and painting, but was obliged at the wish of his parents to study medicine, for which purpose he entered the University of Pisa at seventeen years of age. There the teaching was chiefly based upon Aristotle, and Galileo found it of extremely little value, as appears from his later writings. This seems to have aroused in him a general objection to the study of ancient authors, for in spite of the advice of his father, he only determined quite late, in his twentieth year, to study Euclid. But only this start was needed to show him how to use the most fundamental gifts which he possessed. Euclid, and soon also Archimedes, then engaged his entire attention.

There can be no doubt that Euclid led him to a clear view of the nature of all strict science, and the impression produced upon him by the works of Archimedes was very powerful. 'Only too clearly,' he says himself at the commencement of his earliest essay, 'do we see from these works how greatly all other minds are inferior to Archimedes, and how little hope anyone can have of inventing anything approaching his creations.' Also, his own first appearance in literature is connected with Archimedes; he produced two studies on floating bodies and the centre of gravity, which were at first circulated in manuscript, and only later printed. At that time he was making numerous determinations of specific gravity by determining the weight of bodies immersed in liquids.

He does not appear to have arrived at a regular termination of his study in Pisa, but his well-wishers recognised his unusual abilities, and obtained for him the possibility of earning his living in Florence by giving mathematical instruction; ultimately he was given, at the age of twenty-five, a professorship of mathematics at Pisa. Here he began to concern himself with the problems of motion, by the solution of which he later showed his greatness. But success did not by any means come to him very quickly in this work;

while all the problems of motion which he dealt with – simple free fall, vertical and inclined projection with its peculiar path, the pendulum – were no doubt taken up during his time in Pisa, their true solution was only found later, and in the connection between all these problems. At the various intermediate stages of his knowledge, he met with ideas already put forward by Hipparchus, but still incomplete, and from time to time he imagined that he had conquered the difficulties, but only to find that he was wrong. This gradual development is of high interest;[1] we cannot follow it here in detail, but we may point to this case as an example of the erroneous assumption of peculiar inborn gifts.[2]

The key to all this was a knowledge of the law of inertia, and of the fact that the velocities attained under the influence of a force must be regarded as dependent upon time and not upon distance; that is to say, it was necessary to form the concept of acceleration in its quantitative aspect, as familiar to us to-day. Publication followed late in the famous *Discorsi* (*Dialogues*); in the meantime other discoveries engaged his attention. There can be no doubt that observations on falling and projected bodies and pendulums already attracted his attention during his stay in Pisa. But we have very little accurate information of the details of this, much less

---

[1] The best description of Galileo's scientific life based upon recent research will be found in E. Wohlwill's *Galilei und sein Kampf für die Kopernikanische Lehre* (1909 and 1926). See also *Galileo*, by W. W. Bryant, in the S.P.C.K. series 'Pioneers of Progress' (1918); also J. J. Fahie, *Galileo: His Life and Work* (London, 1903).

[2] The phrase so often met with of the 'intuition of genius' as peculiar to great investigators, is erroneous; only persons without understanding can have originated it. It may of course appear to the outsider as if important discoveries and great steps forward are the outcome of a special gift for guessing the secrets of nature. But enough scientific investigation, observation and thought have always been performed by others previously, and thus sufficient insight into the actual behaviour of nature is already available, and only requires to be worked upon further. This is often a comparatively easy matter for someone who has hitherto taken no part, and approaches the matter with a fresh mind. The earliest workers in science found little store of such knowledge; they had to make their own provision for it.

than was generally assumed for a long time to be the case.

The Pisa professorship, which was not a very good one, was exchanged by Galileo after three years for a better one at Padua, where he remained for eighteen years, which were certainly the most fruitful and happy of his life. Here he further developed his investigations on the phenomena of motion. But a second science was also founded by him, that of the strength of materials. These investigations also were published very late, in *Dialogues concerning Two New Sciences*.[1] Galileo developed the fundamental idea of tensile strength, combined this with the theory of the lever, and so deduced laws concerning the breaking strength of rods, prisms, and cylinders of various dimensions loaded in various ways. He recognised and explained the advantages of special forms of support differing from the prism in respect of greater strength combined with less weight, and the advantages of the tubular form as compared with the solid cylinder. These form the first beginning of a practical theory of the strength of materials, and also of the theory of elasticity. A noteworthy part of this work is the way in which Galileo specially treats of the nature of breaking strength itself; that is, the cause of the solidity of solid bodies. One cause he sees in the 'Power of the Vacuum,' which presses bodies together from outside, whereby he forms a perfectly correct conception of the pressure of the air at a time when anything of the kind was completely unknown. At that time all visible effects of this pressure, as for example the rise of water by the suction of a pump, were ascribed to an indefinite hatred, on the part of nature, for empty space, the so-called *horror vacui*. Galileo measures the 'Power of the Vacuum' by a special experiment with a cylinder and piston, and further, by the limiting height of eighteen ells (thirty feet) beyond which a perfect pump cannot suck the water from a well; this was a matter of observation. From the force

[1] Eng. trans. by H. Crew and A. de Salvis, 1914.

thus calculated he concludes, knowing the specific gravity of copper to be 9, that the highest column of copper which could be supported by the power of the vacuum would have a height of only $18 \div 9$, that is, 2 ells. This brought him almost to the point of Toricelli's later experiment with the mercury column. He also measured for the first time, correctly as far as the order of magnitude, the specific gravity of air, for which purpose, since he did not possess an air-pump, he devised special methods.[1]

Inasmuch as he then found the tensile strength of marble, for example, to be four times as great as the power of the vacuum alone, he was obliged to assume the existence of a peculiar force apart from the vacuum, this force being different in the case of different bodies, and being the internal cause of their strength. The nature of this remained undetermined; the limits of knowledge at that time were reached even for Galileo. For a moment he allows his Sagredo (who is himself) to say in the *Discorsi* that he 'cannot understand how the coherence of the smallest particles, down to the very smallest of the same matter' can bring about the tensile strength; the attempt is then made, 'not as absolute truth, but as a still undigested idea,' to manage with the power of the vacuum alone, perhaps by some roundabout means. This shows how little reason was given by the state of knowledge at that time to assume the existence of peculiar attractive forces acting between portions of matter. Such a conception needed Newton's discovery of universal gravitation, almost one hundred years later. Galileo's discovery that the strength of all bodies is overcome by gravity, as soon as their dimensions become sufficiently great, is also very noteworthy; bridges, houses, trees, and animals, when constructed of the same materials and in the same way, therefore

---

[1] What Aristotle put forward as the weighing of air, namely comparison of the weight of a bladder blown up and collapsed, can only be taken, on account of the buoyancy of the air, as the apparent confirmation of a previously formed opinion, as is often the case with Aristotle.

cannot exceed certain dimensions, since they would other-wise collapse under their own weight.

In the 'First Day' of the *Dialogues* a great deal concerning sound vibrations and sounds is made clear in connection with Pythagoras and previous remarks of musicians; thus the deter-mination of the pitch of a note by its frequency, the relation between the frequencies of harmonious and inharmonious intervals, such as the octave, fifth, third, and second. The dependence of the frequencies of strings upon their weight and upon the tensile force is also completely understood. We find a thorough treatment of the phenomena of resonance which later became of such importance for the whole of physics, and this perfect and exhaustive clarity at so early a period astonishes us.

At that time Galileo was also busy with the construction of thermometers, which were the first appliances enabling 'warm' and 'cold' to be defined quantitatively. By observ-ing the expansion with heat of air, water, and alcohol, he strove to attain to a quantitative grasp of thermal phenomena, and here also he took the path which, in full accordance with the instructions given by Pythagoras, has always been the necessary condition of progress.

Galileo's first discoveries concerning the heavens were also made in Padua, and in their comprehensive importance these are fully on a par with his discoveries on motion. His eyes gazed upon things never seen hitherto, and completely unbelievable, as he turned a telescope towards the sky for the first time. This new appliance, compounded of spectacle glasses, had appeared in Holland, where the grinding of glass lenses for use as spectacles had already been practised for three hundred years, and with very great skill.[1]

The tube with lenses very soon became widely known by its astonishing powers with respect to objects on the earth;

[1] Evidence of much earlier use of lenses of glass and rock crystal, even in antiquity, exists.

Galileo re-invented it immediately he heard of it, and himself made far better telescopes than those which were soon obtainable everywhere, even at fairs. He later on also succeeded in making the transition to the microscope.

Already in the year 1609, Galileo directed his telescope first towards the moon, and recognised, in the mountains casting their shadows into the valleys, the evident similarity of this heavenly body with the earth. He then discovered the almost incredible multitude of fixed stars which had hitherto remained hidden from the naked eye, and which, when seen through the telescope, thickly cover many parts of the heavens; the luminous cloud of the Milky Way was resolved into vast numbers of stars. The enormous extent of celestial space, already recognised by Copernicus and Tycho, now appears to be occupied, to an extent hitherto undreamed of, by suns. But still more than by these innumerable multitudes, Galileo was struck by some stars only recognisable through the telescope, which he found near the planet Jupiter : the four moons of Jupiter. He saw them for the first time in the night of the 17th January, 1610. On the following night he recognised them correctly as moons of Jupiter, since they had changed their position markedly, and nevertheless had remained in one line close to the ecliptic; obviously they were revolving around Jupiter. A secret of the heavens had been revealed. The earth with its moon suddenly appeared quite clearly as having no other rank in cosmic space than that also held by the planet Jupiter; each planet had an equal claim to be regarded as a world in itself, while at the same time it revolved around the sun. The mode of motion ascribed by Copernicus to the earth moon, its revolution around the earth combined with the latter's revolution around the sun, without any means of support, had now to be recognised as possible, however difficult it might be to understand. For anyone who wished could look through a telescope at night and see the reality of such motion; and not merely of one, but actually of four moons, in the case of Jupiter. At the same time the most definite evidence was obtained that there are heavenly bodies moving in closed paths which have not the earth for their centre. Surely never before, and but seldom since, has a single person been privileged to experience such a wealth of completely new observation and thought as Galileo must have done at the time of his discoveries with the telescope and their examination in detail. He himself writes to the Florentine Court: 'I am filled with infinite astonishment and also infinite gratitude to God that it has pleased Him to make me alone the first observer of such wonderful things, which have been hidden in all past centuries.'

These discoveries also led to the fulfilment of Galileo's wish for a position in which he would not be pledged to any definite lectures: it was offered to him by Prince Cosimo of Medici in Florence, and he lived there from September 1610 onwards. His activity as a lecturer in Padua could give him no more permanent satisfaction than his earlier work of the kind in Pisa; at all universities of that time it was absolutely necessary to teach Aristotle. Galileo appears to have developed a certain amount of humour in the process, but it would be wrong to assume that he lectured on his own studies and discoveries. This would have aroused great opposition; for even Copernicus was still looked upon as someone to be laughed at and hissed off the stage. The most prominent professors of philosophy and physics at the Italian universities uttered violent protests against the *Sidereus nuntius* ('The Sidereal Messenger') in which Galileo published in print his new discoveries; they confuted what had been seen through the telescope by means of logic; wild and ill-founded contradictions, spread irresponsibly, formed the main element. Galileo, to meet this, undertook on the one hand the construction of good telescopes in large numbers, and sent them to influential personalities for their use; on the other

TYCHO BRAHE

SIMON STEVIN

hand he held three public lectures before a large body of hearers. But his clear discourse and his good instruments were powerless against his opponents, who persisted in refusing to see the satellites of Jupiter, or even the telescope. Among his colleagues, his young contemporary in Prague was almost alone in supporting him with comprehending and full admiration. Some educated laymen, however, whose minds were unprejudiced by learning, received a very great impression from Galileo's *Messenger*, which was widely circulated.

Galileo himself continued for the time his observations by means of the telescope. His observations of the moons of Jupiter kept him very much occupied; but the determination of their periods of revolution gave great difficulty, since the telescope was not yet adapted for good measurement. He soon recognised that the inmost moon moved fastest and the outermost slowest; his determinations of their periods followed later. He then discovered the peculiar form of Saturn, which appeared like a body divided into three parts, this being the first indication of the ring system of this planet. He further discovered the phases of Venus, which were like those of the moon; and this proved the planet to be dark in itself and only illuminated by the sun, a fact at that time not merely unknown, but regarded as completely untenable. At the same time these phases also proved clearly that Venus revolves around the sun. When stars, dark like the earth, were seen through the telescope, shining in the heavens and moving, it became less and less incredible that the dark earth itself might be also a planet, a star, circling around the sun. Finally, Galileo's observations of the spots and protuberances of the sun showed that this mighty sphere also rotates. For the first time, the motion ascribed to the earth could be directly seen in a heavenly body.

All this raised Copernicus' picture of the world to such a high degree of certainty for Galileo that he gradually abandoned the reserve which he had hitherto maintained, and

Ds

commenced to expound publicly his view of the earth, the
planets, and the fixed stars.   While at this time his discoveries in the heavens were generally recognised, as good telescopes came into existence, a new opposition was awakened
by his plain statements concerning the conclusions which he
drew from these discoveries.   An earth revolving on its axis
was a horror not merely to all sympathisers with Aristotle,
but in particular to the powers of the Church.   The better
the reasons brought forward by Galileo in favour of rotation
and against all objections, the greater the bitterness of his
opponents.   Jesuits and Pope began to fear for their mastery,
when doctrines were spread which contradicted their own.
On the 25th February, 1615, the Roman Inquisition began to
busy itself with Galileo, and for a period of almost thirty
years which followed until his death they never ceased to pay
attention to him, and, indeed, persecuted him with increasing
severity.   For twenty years Galileo remained full of hope
that he would succeed in convincing learned Jesuits, Car-
dinals, and even the Pope of the accuracy of his views.   He
did not realise that his opponents had neither the desire nor
the power to follow him seriously, and that the proceedings of
the Inquisition had nothing whatever to do with the formation of a judgment depending upon the greatest obtainable
knowledge of reality, but merely started from a fixed intention of maintaining the opinion of the Church.   Hence in
1616 Galileo's writings were forbidden, and he was warned
to give up his erroneous opinions under threat of imprisonment.   He submitted.

Later developments showed that his submission was just
as purely apparent as the investigation of his opinions by his
judges.   Galileo behaved to the end towards the power of the
Church differently than did Giordano Bruno, who, sixteen
years before, had turned away with horror, even at the
stake, from a Church which opposed the recognition of
truth; he refused the cross held to him at the last by way of

reconciliation. Bruno finished his earthly labours at the stake. Galileo's work was still to be done; what he had to give the world was still stored up in him; his two most important works were not yet written, but only in preparation in many respects; he was able to preserve from loss discoveries that were to form the foundation of all further knowledge of nature on this planet, and through which the final victory of truth was secured. The most careful investigator of all documents relating to Galileo[1] says concerning these dateful hours in Galileo's life: 'No writing by his hand has preserved for us a memory of those hours of deepest agitation and inward struggle, in which we must believe if we have faith in the human heart. When we hear from him again after a short pause, all is over, the great transformation has taken place.' Galileo was still free, and the protection of his Grand Duke proved to be not quite ineffective. His activity in the years following is known to us exclusively from letters, which allow of no doubt that the limits set by the Inquisition were not too closely observed in his intimate relations.

In the year 1624, the choice of a new Pope brought fresh hope to Galileo. He journeyed to Rome and pleaded warmly with the Pope for the truth of the doctrines of Copernicus, in order to obtain a revocation of the judgment of the Inquisition. He also spoke in the sense of the Church, and emphasised the fact that the matter would reflect upon the scientific insight of the Catholics. He received a gracious hearing in six interviews, which were without any result. Galileo then completed one of his works that had been long in preparation, *Dialogues about the Two Great Systems of the World*, in which he stated clearly the case for the Copernican system as against every imaginable objection, and supported it with a series of special proofs from experience. Of the three characters of these dialogues, the advocate's rôle is

[1] E. Wohlwill. See the work already quoted, particularly vol. I, p. 630 (1909).

played chiefly by Salviati, who in a manner peculiarly Galileo's own, combines a respectful gentleness towards his opponent with compelling power of argument. Simplicio is the learned man of the Aristotelean school without however being narrow-minded, as so many living opponents of Galileo were; he suffers in the dialogue as many defeats as he puts forward arguments. The third person, Sagredo, is a cheerful and kindly character and witness, who is delighted to gain new knowledge; he exhibits the feeling of a prisoner set free, which was echoed by many contemporary readers of the *Dialogues* in letters which are still preserved. Galileo here gathers together every result of his life's work bearing upon the question of the earth's movements. Much of this has already been referred to above. But, as of fundamental importance in these dialogues, we also have the knowledge of the undisturbed super-position of motions of different origin, and the law of inertia. Though all this is very simple, no one had been clear about it before Galileo, not even Kepler, and a mass of objections to the motion of the earth had arisen from this want of clarity.

But the abolition of these was not the only gain which Galileo brought; we owe to him the actual foundation of the whole science of motion, and hence of the physics of matter. The *Dialogues* make use of experimental proof in the form of observations which many people could have made before, but which had not been grasped with sufficient clearness, certainly for the most part through the hindrance produced by learning Aristotle. Thus, for example, we have the fact that the ball falling from the hand of a rider in full gallop, does not remain behind him but retains the speed of the rider, even when it falls at the same time. Or the experience which we have upon ships sailing no matter how fast, that all processes of motion take place on them in exactly the same way as on stationary ships, so that if experiments are made in a closed space, upon a ship, it is impossible to tell whether the

GALILEO GALILEI

JOHANNES KEPLER

ship is in motion or at rest. Also the peculiar effects which are noticed when the ship starts or stops, and the law of inertia comes into action, are dealt with; for example, water in a vessel upon a ship. All this is clearly put forward.

The dialogue received the papal *imprimatur*, although certain alterations had to be made, particularly at the beginning and end, and Galileo agreed to these. The publication aroused great enthusiasm on one side but fierce hatred on the other. The further Galileo progressed in knowledge, the sharper became the opposition between the spirits of light and the spirits of darkness; but the first were, as always, in the minority, and the power was in the hands of the latter. Power of this kind desires unlimited rule for force, for it feels that it is opposed to the progress of reality, and hence that the smallest advance in the ordinary sense of reality can become fatal to it. But power of this kind easily spreads by means of intrigue, and finds the means to build up great majorities for itself out of weak minds.

The Jesuits managed this with great skill in their opposition to Galileo; with suppressed fury they saw themselves made ridiculous, since the complete defeats suffered by Simplicio in the *Dialogues* were also their defeats. They did not rest until they had convinced the Pope that Simplicio could be meant for none other than themselves. On the 1st October, 1632, the sixty-nine year old Galileo was again cited before the Inquisition in Rome. After a journey which itself was a severe trial to him, the hearing took place, which ended in the formal rejection by Galileo of the doctrine of the moving earth, according to the predetermined plan of the Pope ! It is no longer possible to obtain the truth concerning Galileo's statements, since the documents of the hearing, which were only open to public inspection 250 years later, appear to give rise to the suspicion of interpolations and erasures.[1] There is nowhere any question of testing the proofs of Galileo's

[1] See the work of Wohlwill already referred to on p. 27, footnote.

statements, or of a consideration of reason in his favour: but instruments of torture appear to have been ready to hand during his examination. In any case, Galileo left Rome as the prisoner of the Inquisition, and remained so until his death. After some time, however, he was allowed to live in his country house Arcetri in the neighbourhood of Florence; but the Florentine Inquisitor kept him under continual close observation. He was almost entirely confined to the house and could only receive visitors who were approved by his gaolers.

The public effect was that of complete paralysis. No protest on the part of science, even from the universities, is recorded by history.

During his imprisonment, Galileo wrote his second and last chief work, the *Conversations concerning Two New Sciences* (*Discorsi*, of which we have already given a partial account above). It contains all his investigations of the laws of motion, which had occupied him almost without interruption since his youth, and which made him the founder of this science. He gives for the first time a complete statement of, and argument for, the results, and develops a large number of new conclusions. Here we find the law of free fall, of falling on the inclined plane with corresponding experiments – the times being measured by the flow of water, since there were still no good clocks – and further, detailed treatment of the motion of projectiles, the laws of the pendulum, and the consideration of many other important phenomena of motion. The printing of the work was prevented by the prohibition of the Inquisition in all Catholic countries; but by friendly endeavours, the publication was arranged through a Dutch bookseller, though the fact that this was Galileo's desire had to be kept secret.

When the work appeared, Galileo was 74 years of age. A year before that he had already become blind after a long and painful disease of the eyes. Petitions to the Pope to allow him to exchange his lonely and out of the way dwelling place, where every kind of treatment was rendered more difficult, for his home in Florence, were rejected with rough words and threats of a Roman prison. Finally, the Inquisitor visited him accompanied by a doctor; he found him blind beyond recovery and in so wretched a condition, that he is said to have looked more like a corpse than a living man. A short journey to Florence for the purpose of medical treatment was then allowed, with the order that he should not leave the house on pain of lifelong imprisonment, and should speak to no one whatever concerning the forbidden doctrine of the earth's motion. He returned to Arcetri and died there on the 8th January, 1642, at the age of 78. He only received a worthy tomb ninety-five years later.

The event proved the truth of what Galileo had written in a letter after his first contact with the Inquisition: 'I believe that there is no greater hatred in the whole world, than that of ignorance for knowledge.' The progress of knowledge has nevertheless proved incessant. But for the very reason that knowledge continually crosses new frontiers, and that the spirits of darkness in power, whose mastery is founded upon ignorance, are always able again to become dangerous, that hatred never dies; it simply attaches itself to new objects. It must always be turned against the spirits of light when they bring new knowledge, and it will always prove devastating, as long as the mass of the ignorant, who are available in its service, is not reduced.

JOHANNES KEPLER
1571–1630

KEPLER, who was born later and died earlier than Galileo, stood towards his older contemporary in much the same relation – if such comparisons can ever be of value – as

Schiller to Goethe or Liszt to Richard Wagner. But Kepler and Galileo never met; only their correspondence tells us that they were two great men who, surrounded by the incomprehending, were able to understand one another as individuals and appreciate one another's merits. This is particularly true in a high degree for Kepler; for he had a very fine and delicate mind, and was always ready with admiration and recognition for contemporaries and predecessors; in which respect, quite by way of contrast to the custom of his time, he sometimes even went too far in his writing.[1]

Kepler was the great investigator in optics, as well as of the ideas of the Copernican planetary system. In the first capacity he was the founder of geometrical optics, that is to say, all knowledge based upon the rectilinear propagation, reflection, and refraction of light. In the second capacity he was the discoverer of the three planetary laws, which were fundamental for all further progress in the mechanics of the heavens. While it was still forbidden even to think of the earth as performing its orbit, Kepler had already discovered the more exact form of its path around the sun, together with the paths of the other planets, their mode of motion in their orbits, and the relations of the periods of revolution of the different planets to one another.

The three well-known laws of Kepler are as follows: (1) The paths of the planets are ellipses, with the sun at one focus ; (2) The line drawn from the sun to the planet sweeps over equal areas in equal times ; (3) the squares of the periods of revolution of the different planets are to one another as the cubes of their mean distances from the sun. This refinement of knowledge was taken by Kepler from Tycho Brahe's fine observations with an expenditure of labour and acuteness which had never been known before,

[1] Thus for example in the case of the writer and self-advertiser Porta, who at that time attained a certain degree of importance; indeed. Kepler generally allows even Aristotle too great significance.

and for long had no equal. Nothing was added to this body of knowledge for seventy years, until Newton with his discovery of universal gravitation, based upon Kepler's work, made a further advance of comprehensive importance into the unknown.

Kepler was free from the persecution which Galileo had suffered as a result of his knowledge, since Luther's life-work saved him; he was a Protestant.

Nevertheless, his life in the land of the thirty years war, which began in his forty-ninth year, was full of misfortune and need, and just because he was a Protestant he was exposed to much persecution.

He came from a family originally noble, but which had fallen to poverty. He was born on the 27th December, 1571, at Weil, a town in Wurtemberg,[1] the father was mainly absent as a soldier in foreign countries, and Johannes had to help on the farm at home, but was also able to go to school. When it was seen that he was too delicate constitutionally for anything else, he was allowed to study at the University of Tübingen, which he attended from his sixteenth year, and where he actually received a stipend on account of his good work at school. He was fortunate enough to find there a splendid teacher of mathematics in the person of Mästlin, who also taught him the work of Copernicus, though even in Tübingen at that time, they hardly dared to recognise it too openly.[2] After two years he obtained his degree with distinction. He then proceeded to study theology. But before this study was finished, the opportunity came to him to take a professorship of mathematics at the evangelical school in Graz, and this opportunity he took. At the age of 23 he began his work there, but this also inevitably included making

---

[1] The town possesses, since the year 1870, a beautiful and worthy statue of Kepler.

[2] Even Luther and Melancthon were its opponents, prejudiced by the Old Testament, but they did not make use of the power of the stake to enforce their views.

an almanack each year, with prophecies of the weather and of political events. Astrology of this kind had unfortunately to be Kepler's main source of income to the end of his life. He calls it 'the foolish and disreputable daughter of astronomy, without which the wise old mother would starve.'[1]

In the Graz period we have Kepler's first attempts, never afterwards abandoned, to discover regularities in the planetary system. At Graz, also, he founded a family; but quiet happiness was not to be his lot. The persecution of Protestants, which soon commenced, destroyed all security of property; and finally, all those serving the Protestant Churches and Schools were banished from the land, as a Hapsburg (Grand Duke Ferdinand) educated by the Jesuits came to the throne. Kepler was the only one allowed to return. His rare gifts were highly valued; but it soon appeared that he was expected to enter the Roman Church. Since his conscience would not allow him to do this, and since he frequently and openly made the fact known, all kinds of disciplinary measures began. In this time of oppression, two of his children had died; about the year 1600, threats of imprisonment and torture were made. An invitation by Tycho Brahe at last gave him the desired opportunity of leaving Graz, which had become unbearable for him. At that time, Tycho had just moved to Prague, and Kepler became his assistant there, and after his early death, his

---

[1] In a special essay ('Warnung,' to be found in the 1858 edition of his collected works, vol. I, page 547) Kepler defends himself with great seriousness from misinterpretation of the attention he gave to this 'bastard of science,' and warns against confusing 'star-gazing superstition' with highly intelligent astronomy.' And on another occasion (in a letter, *ib.* vol. 8, page 811) he says: 'Just as if the works of God were otherwise not worthy to be looked upon and to form the subject of our calculations, they are supposed to have some kind of meaning directed to such silly people of peculiar disposition.' Nevertheless, his firm adherence to a connection between the worlds of spirit and matter (a connection still undiscovered), and therefore a connection between human fate and the motion of the heavenly bodies, rendered even this work less difficult. We may compare on this point the fourth book of his *Harmonices Mundi.*

successor, with the title " Imperial Mathematician." Unfortunately, however, the effects of political confusion in Germany, from which he never succeeded in escaping, followed him also to Prague. His salary, nominally ample, was only paid rarely or in part, so that his domestic life was very uncomfortable, the more so when later the war came to his neighbourhood, and epidemics broke out.

Nevertheless, it was just in the eleven years of his stay in Prague that Kepler did the main part of his life's work. He became the scientific heir of Tycho, taking over the extensive records of Tycho's fine measurements, extending over many years, of the positions of the planets, together with the task of making scientific use of these observations, and of finishing the Rudolfine Tables, which were to give the positions of the planets at future times. Kepler attacked this task in the most thorough manner possible; he wished above all to find the true paths of the planets from Tycho's observations, with the greatest possible accuracy which the observations permitted. He soon found that the circular paths performed at constant speed, which alone had been considered hitherto, would not suffice. Further advance was very difficult, for the planetary motions are observed from a moving position, namely the planet Earth, of which the exact path and mode of motion were just as unknown as those of the other planets.

Kepler expended six years of tireless labour upon the path of Mars alone. After all kinds of eccentric paths around the sun for Mars and the Earth had been tried, whereby the diameters of the paths had also to be altered by trial, and after all this had not led to the desired goal, various non-uniform motions in the circles were tried one after another. One of these appeared satisfactory, namely the one in which the line joining sun and planet swept over equal surfaces in equal times. With this the substance of what was afterwards called the second law was discovered, but Kepler's delight over this discovery was of short duration. The agreement with reality as known through Tycho's observations was certainly better than hitherto, but it was not complete; small discrepancies remained. Kepler was not the man to be satisfied with partial success; it also fell to him to give us a first great example in the history of science, of how discrepancies between calculation and observation are to be judged. The departures were small, but – and this was the essential point – they were nevertheless greater than the possible errors of Tycho's observation. As soon as a case of this kind occurs, the conclusion must be drawn that an unknown factor is concerned. Kepler now began to seek the unknown element in the shape of the planetary orbits. Hitherto only the circle had been seriously considered as the path of permanent heavenly bodies; the comets however had already betrayed the fact that other paths are possible. The departure from the circular path could not be great in the case of Mars; Kepler tried, one after another, all sorts of other closed paths of different curvature. The ellipse was at that time no more likely than any other possibility; but only elliptical orbits with the sun at the focus satisfied Tycho's observations; and this they did completely when the law of equal surfaces for the radius was taken into account. With this the problem of the path of Mars had been conquered after six years, and the two first laws had been found, which now could be taken to hold for the other planets as well as for Mars and the Earth. The publication followed in his *New Astronomy founded on True Causes* (1609).

At the same time, Kepler was busy with optical investigations. Galileo's *Messenger*, which had just appeared with the successes of the first telescopic observations of the heavens, led him to investigate closely the path of the light rays in the telescope, and further, the behaviour of light rays generally. The results are collected in two works (*Paralipomena* and

EVANGELISTA TORICELLI

WILLEBRORD SNELL.

René Descartes

Blaise Pascal

Dioptrics) which collected all that had been known from the time of Euclid, and made quite clear how, for example, the pinhole image (Camera obscura) is formed, but then went very much further. Thus we have there for the first time the decrease in the strength of light according to the inverse square law in the case of free propagation from a small source, the correct theory and calculation of mirror images, the refraction of light and the theory of lenses; and also the explanation of the path of light in the eye, and thus all the essential facts of sight, as far as light itself is concerned, including the correct explanation of stereoscopic vision with two eyes. It was true that Kepler still made use of a law of refraction which was only approximate, but this did not prevent him from recognising total reflection and grasping it directly. Finally, also, he was the first to state the principle of the telescope with two convex lenses, which is to-day used exclusively and quite generally for astronomical work in place of Galileo's telescope (which had a combination of a convex and a concave lens), and also serves many other purposes, being called the 'Kepler' telescope. The present-day telescopic objective is also given.

In the meantime, Prague had become less and less endurable for Kepler. The Imperial Mathematician had no money to live upon, and nevertheless the Emperor (Rudolf II) would not allow his Kepler to escape. A further misfortune was the death of his wife and one child. So Kepler had gradually brought himself to face the need for leaving the place he loved so well, and looked around in his own home country. The Court of Stuttgart would have been glad to have him at the University; but Kepler's independent cast of mind made him objectionable to the ecclesiastical authorities, although these were Lutheran.[1] So in 1612, he went to a school at

---

[1] In this we may compare the fate of Böhme in Silesia, at the same period, where the new enthusiasm awakened by Luther had taken a still more unfortunate and erroneous direction, even among his own followers.

Linz, where he remained fourteen years. Here he married
for the second time; but in this period also he suffered much,
for example, the witch trial in which evilly-disposed persons
had involved his mother. He hastened to her and saved her
by an effective defence when she was about to be tortured.
His work in Linz was his most comprehensive book, the
Harmonices Mundi, and he also completed there the Rudolfine
Tables. The latter then remained for a century the foundation of all planetary calculations. Of the five books of the
Harmonices (which appeared in 1619), the fifth has become
of undying importance by its announcement of the third law
of planetary motion, which here appears joined with the two
first laws. Kepler discovered it from the figures given
by Copernicus and Tycho for their observations; he describes
it himself as being the result of 17 years of work. His
own high valuation of this part of the *Harmonices* is shown
by the following sentence in the preface: 'I shall cast the
die and write a book, whether for the present day or for
posterity – it is the same to me. Let it await its readers for a
century – God himself has waited for his interpreter for six
thousand years.'

High enthusiasm for his work and its results enable him
again and again to rise confidently above the confusion of his
external life.

The remaining books of the *Harmonices* deal with
geometry, music, the theory of harmony, and with the connection of the latter, and also of human life, with the
motions of the planets. Here much seems strange to us
to-day; but this feeling is only justified by our having come
to see clearly that the relation between spirit and matter
cannot be of so simple a character as was imagined in Kepler's
time. But Kepler's search for such relations was unjustly
despised; for they must exist, since spirit and matter are
actually connected with one another in the processes of life.
If to-day few traces of this search are to be found in works of

science, the reason is to be found in their comparatively small scope and in the dominance of materialism, from the narrow and dreary outlook of which the spiritual world has completely disappeared. When Kepler says, for example: 'And I have always striven to investigate, with an open mind and by the use of reason, what the nature of spirit may be, and principally whether there may not be a world soul in the heart of the world, which is more deeply connected with the processes of nature . . .' great men of science will still be able to follow him; now perhaps attaching to his words an essentially deeper and preciser meaning.

Kepler's external life, even in Linz, finally became less and less favourable under the stress of the Thirty Years War; five children of his had died there, and his salary ceased completely. Ferdinand II made over to him the 12,000 guilders owing to Wallenstein, who was well-disposed towards Kepler, since the latter had once given him an astrological result of importance. So Kepler moved under Wallenstein's protection to Sagan, where comparative quiet had reigned during the war. But his salary remained unpaid. In order to make a personal appeal for it after Wallenstein's dismissal, which had shortly afterwards occurred, he set out for the Reichstag at Regensburg, but exhausted by the fatigues of the long journey on horseback, he fell seriously ill on his arrival, and died shortly afterwards, on the 15th November, 1630. His epitaph, composed by himself, is as follows:

> *Mensus eram coelos, nunc terrae metior umbras;*
> *Mens coelestis erat, corporis umbra jacet.*

His grave in the churchyard near the fortification of Regensburg was so completely buried later on in the war by the bombardment of the walls, that it can no longer be found.

EVANGELISTA TORICELLI (*1608–1647*)
BLAISE PASCAL (*1623–1662*)

THESE are two successors of Galileo as regards a part of his work, and likewise forerunners of Guericke, as regards the complete elucidation of the connection between atmospheric pressure and the weight of the air.

Both had only a short life.

Toricelli was actually a pupil of Galileo's in his old age, and became his successor as mathematician to the Grand Duke of Tuscany. He was obviously a born experimenter. This is already shown by his remark that small beads of glass, which one can easily melt oneself, form excellent lenses of high magnifying power; and many years later these were still preferred to the compound microscope, which is much more difficult to construct, and were used to make important discoveries in the domain of the smallest forms of life. By means of a series of quantitative experiments he discovered the law, still named after him, of the flow of liquid from openings in thin walls, whereby Galileo's laws of fall proved to be valid in the simplest manner also for liquids. This was the first law ever discovered concerning the motion of liquids (hydrodynamics) as contrasted with the science of hydrostatics founded by Archimedes.

Toricelli's best known achievement was the construction of the apparatus for measuring air-pressure, which was soon, and is still called the barometer. Galileo had already measured this pressure, the 'power of the vacuum,' by means of the height of the water column in the tube of a deep well in Florence, and he had also calculated the much smaller height of a column of copper which could be supported by the same pressure. From this it was not far to the idea of trying a mercury column, which, like the water column, could of itself assume a height corresponding to the pressure to be measured. Toricelli communicated his idea to his friend

Viviani, Galileo's youngest pupil, who then first carried it out in the year 1643, in which Galileo died.

In this way it was possible to show in small dimensions, and visibly by means of a glass tube, what the water column in the well only allowed us to guess: the self-limited and self-adjusting liquid column, which has the same pressure as the atmosphere, and quite by itself leaves an empty space above it; the possibility of such a thing had been for long regarded with much doubt. Toricelli then repeated the experiment with alterations, and made continued observations of the height of the column, whereby he also became the discoverer of the variability of atmospheric pressure. The publication of these discoveries took place only in a letter, but at that time this was not an unusual procedure, and sufficed to spread the new knowledge, inasmuch as it has always remained connected with the name of its discoverer.

Blaise Pascal, who was born at Cleremont in the Auvergne, already showed quite uncommon powers in early childhood. He began to master Euclid at the age of twelve, wrote at eighteen essays on conic sections of permanent value, and also became early acquainted with Galileo's work. As soon as he had learned enough concerning Toricelli's discovery of the pressure of the air, it appeared to him of prime importance to prove by a further series of experiments, that it is actually the weight of the air which exerts the pressure and holds up the mercury column. For this purpose he made a special arrangement of two connected Toricellian tubes, so arranged that one could remove the air supporting the other; he then actually saw the latter fall down completely. When air was readmitted, the column rose again. He also carried out many experiments with apparatus provided with mercury in flexible bags and sunk in siphon tubes under water, and found that the pressure of the overlying water produces exactly the same effects as those observed in Toricelli's tubes,

Es

and hence that these may rightly be ascribed to the pressure resulting from the weight of the air. He immediately wished to carry out a decisive observation of the barometer on a high mountain, since there the less amount of air overhead should result in a correspondingly shorter column of mercury. Since his own neighbourhood gave him no opportunity for such an experiment, he asked his brother-in-law, who lived at Cleremont in the south of France, at the foot of the 1,000-metre Puy-de-Dome, to carry out the experiment on this mountain. He did so with great thoroughness, and found the height of the mercury column diminished by several inches.

Thus in 1648 it was for the first time proved without doubt that in the air-ocean of the earth the pressure diminishes as we go upwards, just as it does in a mass of ordinary liquid; the latter fact, by the way, being already set forth quite clearly in Stevin's writings. The evidence was convincing that it is the weight of the air, already demonstrated by Galileo, which supports Toricelli's mercury column, and not Aristotle's 'horror of a vacuum,' in which many, including at first Pascal, still believed firmly. But there are no grounds for this being different high up than upon the earth. Pascal was astonished at the great difference in the mercury column found by the experiment on the mountain; he then ventured to make experiments also with smaller differences in height, on towers and houses, and found all his expectations confirmed.

Pascal was also the author of extensive writings on the equilibrium of liquids, and the hydraulic press was his invention; but here all he did was to apply in new ways what had already been mainly discovered by Stevin. The religious speculations also, which occupied Pascal's last ten years exclusively, did not lead him beyond ideas already existing at his time.

WILLEBRORD SNELL (SNELLIUS) (*1591-1626*)
RENÉ DESCARTES (CARTESIUS) (*1596-1650*)

HERE we have two unlike contemporaries associated, but they approach one another closely in their achievements as men of science.

Snell was born in Leyden and lived there as professor of mathematics and mechanics, where he was the successor of his father. He made the first measurement of the meridian by the method of triangulation, the only one trustworthy as regards accuracy, and used to-day for every kind of surveying. It also became of importance in determining the unit of length the metre (one ten-millionth part of the earth's quadrant). He further discovered, in about 1620, the important law of the refraction of light. Measurements of the angles of refracted rays on entering water or glass had been made earlier, also by Kepler, but the single law which connects the angle of incidence with the angle of refraction had not been discovered, because attention had not been paid to the angle between the ray and the normal to the surface, called to-day the angle of refraction; the more obvious measurement had been made of the angle between the incident and refracted ray, which angle depends in a very complicated manner upon the angle of incidence. It cannot be doubted that Kepler, who discovered from Tycho's figures the deeply hidden planetary laws, would also have discovered the law of refraction, if he had devoted his whole attention to it; but the approximation used by him was sufficient to provide him with a basis for his extensive discussion of geometrical optics. But as soon as the law was found with exactness, a knowledge of it must, in view of its great simplicity, have spread very rapidly. Hence we cannot regard it as of importance that Snell was prevented by his early death from getting the essay containing this law printed; for Huygens and others had seen the essay, and the former also describes in one of his own works (*Dioptrica*) the law of refraction discovered by Snell, upon which he had also founded the important explanation of refraction in his famous 'Essay on Light.' The same thing here obviously occurred as with Toricelli's letter, only that Snell's name was not mentioned afterwards so prominently as Toricelli's.[1]

René Descartes, born at Lahaye in Touraine, of an old noble family, was educated in a Jesuit school and then destined by his father for the Army. He had a very adventurous life, divided between participation in many campaigns of the Thirty Years War in Dutch, Bavarian and Austrian service, extensive travelling, amusement, and then complete retirement; he died at Stockholm at the age of 54.

His mathematical gifts, which appeared very early, enabled him to become the founder of analytical geometry, the method of representing lines and curves by equations; a method which has become of the highest importance for all scientific work, and which has enabled mathematics to assist in all investigations of processes taking place in space. He was the author of other mathematical achievements as well. His only direct valuable contribution to scientific investigation is probably his thorough calculation of the rainbow. He applied Snell's law of refraction for the first time to the course of the ray in the raindrops, and thus found, from a careful calculation of a thousand ray paths, the particular angle, which could not have been found without the law, according to which the rays reflected once and twice in the drop, emerge parallel, and are thus able to affect the eye, whereby they form the first and second rainbow. The formation of the bow by inward reflection in the drops had, however, already been suggested by the little-known

[1] See on this point, particularly as regards the frequent ascription of the law of refraction to Descartes, Mach's *Principles of Physical Optics* (trans. by J. S. Anderson and A. F. A. Young, 1926).

Archbishop Antonius de Dominis[1] before 1611 as the result of observations of glass balls and spherical bottles filled with water; but the agreement of the angle calculated by Descartes with the actual bow was nevertheless an important proof of the correctness of this explanation, and hence of all knowledge connected with it. The colours of the bow were, however, left unexplained by Descartes, and this success remained for Newton.

Descartes made extensive attempts at the solution of many other further problems, including a new foundation for the mechanics of the heavens. But how little his manner of thought was suited to scientific investigation in a larger way is seen from his condemnation of Galileo, for reasons given in his letters; for example: 'Everything that Galileo says about the philosophy of bodies falling in empty spaces, is built up without any foundation; for he ought first to have determined the nature of weight . . .' To think like this is to put the bridle on the horse of scientific investigation at the wrong end, which all experience has shown to be an unfruitful proceeding. In actual fact, the Galilean horse had still to march on with its head in front for a considerable time, to the discovery, observation and interpretation of new facts, in order to approach the question of the nature of weight. The extent to which Descartes' mind was directed towards grasping the non-material side of the world is seen by his filling space with eddies of an unknown substance, whereby the planets were driven around the sun, and also by his widely known conclusion: *Cogito ergo sum* (I think, therefore I am), whereby in determining his own existence he is not concerned with the existence of his body, nor with impressions on his senses, but simply with the existence of his mind.

[1] Who perished in 1624 in the prison of the Inquisition. His contributions to the theory of the rainbow were also noted by Newton in his *Opticks*.

OTTO GUERICKE
1602–1686

THIS distinguished investigator, who is well-known as the inventor of the air-pump and Burgomaster of Magdeburg, began his dialogues with nature almost without any equipment of learning, in experiments thought out by himself, which in many directions carried him beyond this time, and deep into hitherto unexplored regions. But the mere invention of air-pump has little to do with his right to rank among the great men of science; this depends rather on the way in which he used that instrument to put questions to nature, the answers to which could not possibly have been found without its assistance; and he scarcely ever ceased his efforts until each question was settled as far as was possible at that time. The fact that he had a taste for carrying out his experiments on a somewhat large scale, as with the 'Magdeburg Hemispheres,' made them all the more impressive to his contemporaries, but also resulted in effects, as in the case of the experiments with an electrical machine, which would not have been noticed had the scale been smaller. A very sympathetic trait of character in this man, who almost throughout his life was a leader among his fellow-citizens, was his pleasure in the astonishment of the uninitiated, when he made the effects of nature seen or felt by means of his apparatus; the same apparatus which had also revealed to him marvels which not everyone could appreciate as he could. In this way Guericke was unique of his kind.

The descendant of a highly-esteemed senatorial and patrician family of Magdeburg, young Guericke was destined for the study of law, which he began at the age of fifteen years at the university of Leipzig, continued at Helmstedt and Jena, and finished at the age of twenty-one in Leyden. At the latter university he also had the opportunity of hearing lectures on science and mathematics, which, in the midst of the Thirty Years War then raging, also dealt with war material. Snell might also have been his teacher there.[1] A journey to England and France followed, and he finally returned to his birthplace at the age of 23. Here he soon married, and began to give valuable service in matters connected with the town during these difficult war periods.

During the siege of Magdeburg (1631) he acted as Senator, 'Defence or War-Lord' of the town, and did his duty to the utmost in fortifying and defending it. But its resistance was broken by the superior force of Tilly, and Magdeburg was completely destroyed and burnt, whereby Guericke lost all his property and nearly his life. Having lost even his clothes, all that remained of the rich possessions of the family was a book of family memoirs, which happened at the time to have been sent away on loan. But he was fortunate in being able to save his family by sacrificing his property, for he and they were imprisoned for some time, until he was able to obtain ransom, probably through the Swedish and Protestant friends of Magdeburg. He was now penniless; having entered the service of Gustavus Adolphus, who intended to relieve Magdeburg, he became his quartermaster-general until Magdeburg, after Tilly had been driven off, finally came under Swedish protection and could gradually be rebuilt. Guericke then returned there. He again rendered great services to the town, both as engineer in the building of bridges and fortifications, and also in negotiation, for the confusion of the war continued. He also became an agriculturist and brewer; the completely impoverished town was not even able to secure his means of livelihood either at the time, or in all the years following, as a reward for his continual unselfish services.

In the year 1646 he was made Burgomaster; from that time

[1] In the only detailed biography of Guericke, which is derived from original sources, and written by F. W. Hoffmann (Magdeburg 1874), there is no statement of this.

until 1660, he was almost continuously occupied in the business of the town, and travelled a great deal to Vienna, Prague, Regensburg, to the heads of states, princes, and state assemblies, with the object of regaining the ancient rights of Magdeburg. Nevertheless he again and again found time in this period of his life to devote himself to the prosecution of his desire for deeper knowledge of nature. Above all, he was greatly interested in the uncertainty, much discussed at the time, as to whether it was possible or impossible to produce an empty space, whereby it appears that Toricelli's experiments, now some years old, must have been known among the learned, but not sufficiently in other directions, particularly in Germany, which, at the close of the Thirty Years War (1648), was for the most part laid in ruins, and deprived of all peaceful civil life.

Guericke set about using his best efforts to make possible the production of an empty space whenever desired, if such a thing could be done. He attempted to use for this purpose the ordinary well or fire-engine pump, and since this was made for pumping water, he connected one to a cask full of water and otherwise completely sealed up, in order to see whether on pumping out the water an empty space would remain in the cask. The power necessary to work the pump turned out to be so great that he was first obliged to reinforce all joints and fastenings. When finally three strong men working the pump were actually able to pump out the water, a noise was heard in all parts of the cask, as if the water was boiling furiously, and this lasted until the whole cask was filled with air in place of the water removed. Obviously, wood was not sufficiently airtight for such experiments. When the experiment had been repeated fairly satisfactorily with a cask submerged in water, Guericke had a large copper sphere made, which could be attached to the pump, and he now proceeded to omit the water and pump out the air directly, which was entirely successful. As soon as it was supposed that nearly all the air had been pumped out, 'the sphere of metal was suddenly crushed with a loud report, and to everyone's alarm, just as one crushes a cloth between the fingers, or as if the ball had been thrown down from the top of a tower with a loud crash.' This quite new effect, so instructive as regards the magnitude of the atmospheric pressure, was immediately ascribed by Guericke, quite correctly, to the form of the vessel not having been perfectly spherical. It did not occur when a new and perfect sphere had been made. It was now possible to pump until no more air escaped from the valve of the pump, which could be regarded as a proof of complete evacuation. 'So an empty space was thus reproduced for the second time.' After opening a stop-cock fixed to the sphere, 'the air rushed into the sphere with such force, as if it strove to tear with it a person standing by.' Guericke now proceeded to the 'construction of a special machine for making a vacuum' – the first air-pump. This later underwent many transformations, but had from the beginning the essential characteristics of a good air-pump. Guericke also already recognised the inevitable defects of such pumps: leakage of air past the piston, and dead space, and he made arrangements to diminish these as far as possible. As the space to be exhausted, he generally used a glass vessel with a wide neck attached by a ground joint, and suitable for the introduction of experimental objects. The completeness of the vacuum was tested by allowing water to enter, whereby only small bubbles remained, and these were shown to be mainly derived from the water.

There now followed a long period of experiment with completely new observations, which formed the foundations of many later developments. Above all, the tendency of every mass of air to expand through all space, was recognised. The air-bubbles which adhered to the walls of vessels under water already showed this phenomenon, inasmuch as they always expanded when the air pressure was removed. A pig's bladder with a small quantity of air sealed in it, burst of itself when in a vacuum. It was thus clear that during pumping, the air passes quite by itself from the vessel to be evacuated into the empty pump, in so far as it is not prevented from doing so by the back pressure of the valve. Guericke eliminated this by the employment of a special arrangement for opening the valves. He also replaced the latter by stop-cocks. He convinced himself that the automatic equalisation of pressure also takes place through a long tube, when the pump and the vessel were situated on different storeys of the house. He noticed especially the energy with which the equalisation of pressure takes place under suitable conditions, so that small stones and nuts may be hurled about by the air, and concludes from this that winds and storms can only be caused by differences of pressure in the atmosphere; he was once able to prophesy quite correctly a destructive storm from the particularly low pressure of the air. He also noticed the formation of mist when moist air is expanded, and supposed that there is a causal connection between the reduction in pressure of the atmosphere and the formation of clouds.

The hard clash produced by two masses of water meeting in a vacuum was also very remarkable; 'This could not be produced by anyone in any other way than in an empty vessel of this kind.'[1] Guericke also remarked that after some time air-bubbles always again appeared under water on the walls of the vessel, unless it has been pumped out for a long time previously. Since at that time, and even for many years later, the material character of gases, which the air-pump first taught us to regard as something that could be taken hold of, had never been properly grasped apart from the

[1] Commonly demonstrated by the 'water hammer,' a glass tube containing some water, which is boiled to expel the air, the tube then being sealed. [Trans.]

differences given by the sense of smell, these air-bubbles, as well as the whole atmosphere of the earth, appeared as resulting from the escape of volatile constituents from solid substances. Guericke also states quite correctly, that on account of the volatility of all bodies, empty space can always be obtained only approximately, but that this does not prevent our drawing conclusions with respect to it.

The fact that the atmosphere exerts the great pressure now rendered so obvious, Guericke ascribed entirely to its weight; the 'horror of a vacuum' was abolished completely. Instead, he determined directly the specific gravity of air, which Galileo estimated indirectly, by comparing the weight of the same vessel empty and when filled with air, whereby a particularly convincing proof of the considerable weight of quite moderate amounts of air was given by the visible increase in weight of an evacuated vessel while upon the balance, when air was allowed to stream in. Guericke also recognised that we cannot ascribe a definite specific gravity to air, but that it varies with both pressure and temperature. The earth's atmosphere is thus held together by its own weight, and on the surface of the earth it is pressed together to its own density. The vast space between the heavenly bodies Guericke already recognised as empty of all matter.

Through Guericke's investigations, the air became for us for the first time an object which could be grasped; a body which, like solid and liquid bodies, could be introduced into and removed from a space at will. It also became possible to determine by direct observation how space filled with air differs from empty space. Guericke made use of this in two particularly important matters; he first considered the propagation of light and sound in empty space. He points out that empty space, contrary to the opinion at that time, does not prevent light passing through it, since objects contained in it are seen. But his experiments with sound gave a different result. A clockwork which continually struck a bell was hung up by a thread in a glass vessel which could be evacuated; the sound of the bell ceased to be heard after the air had been sufficiently pumped out. But noises were frequently observed to escape from a vacuum, which somewhat confused Guericke; he appears, since these experiments were not continued sufficiently, to have been deceived by the conduction of sound through solid bodies.

The burning of a candle, and the life of animals, failed already in a space from which the air had been only partially removed, and Guericke immediately drew the conclusions that fire takes up something from the air which it needs to maintain itself, and hence that it consumes air; a fact, by the way, also stated by Leonardo. He followed this up in special experiments, using a quantity of air enclosed over water, and found that a burning candle used up at least one-tenth of the air before it went out.

A large number of experiments were made by Guericke with very expensive apparatus, not so much for the sake of the investigation itself, but rather in order to produce an effect on his contemporaries with what he had discovered. In this category are arrangements which caused great force to be exerted, for example a very large copper cylinder with pistons, which, when suddenly connected with a space previously evacuated, was able to overthrow twenty, thirty, or even fifty men, who were pulling on the pistons by ropes. The effective force is given by Guericke from the air pressure as calculated by him (a water column of twenty Magdeburg ells or ten metres) correctly in pounds; but he also tested it by weights on a balance, and states how one can easily calculate the total weight of the air on the earth, if we know the total surface of the earth.

The best known of these arrangements was the 'Magdeburg hemispheres,' the smaller of which, having a diameter of three-quarters of a Magdeburg ell, could only with difficulty be separated by a team of eight horses on either side,

whereupon a loud report resulted; whereas the simple opening of a stop-cock caused them to fall apart of themselves.
The large pair, having a diameter of one ell, was calculated
by Guericke to require two teams of twenty-four horses.
These experiments were shown by Guericke in Regensburg
at the Reichstag in 1654, first to a small circle, and then by
request to the Emperor and the princes. The news of his
discovery then spread rapidly, and particularly the knowledge
and conviction of the magnitude of the atmospheric pressure,
if much had not, as is probable, already become known
already from Magdeburg. Guericke contributed to this by
handing over the apparatus he had brought with him to
some influential personalities who desired it. Part of it then
came to the university of Würzburg, and from there, and
later by Boyle in England, a printed account was published
of Guericke's original experiments, and also of some variations, whereby his priority appears to have been suppressed
to some extent.[1] This caused Guericke himself to write a
book *De Vacuo Spatio* ('New Magdeburg Experiments on
Empty Space'); it was finished in 1663, and finally printed in
1672.

Guericke also engaged in other investigations. He discovered electrical repulsion; hitherto only attractive forces
had been known. For this purpose he made use, in place of
small pieces of resin excited by friction, of a large cast ball of
sulphur, which could be turned about an axis, whereby a
beginning was made of the later frictional electrical machine.
Guericke's original machine, an example of which was
early received by Leibniz, produced electric sparks for the
first time (1672), a fact reported by Leibniz in a letter to
Guericke.

Guericke's old age was clouded by remarkable ingratitude

[1] A striking sign of this is found in Huygens' famous essay on Light
(1678), where Boyle is described as the inventor of the air-pump, though
Boyle himself mentions Guericke as the originator.

on the part of his fellow-citizens; his salary and income of Burgomaster were often withheld, while he was still not allowed to rest even when over seventy-four. There was clearly no one of like skill and devotion available to carry on the continual difficult negotiations which were needed as a result of the Thirty Years War. In any case, Guericke had done his utmost all his life for his native town, obviously from a pure sense of duty to the community to which his father had belonged. That he did not succeed in obtaining the rights of a free town of the Empire for Magdeburg after the war, was not his fault, as the documents prove.[1] He had certainly established the best possible connections with the Hapsburg Emperors (Ferdinand III, Leopold II), who gave him an hereditary title of nobility in 1666, but nevertheless finally decided against Magdeburg; he was likewise highly regarded by the great Kurfürst Frederick William. He died at the age of eighty-four at the home of his son in Hamburg, who was there the 'Kurfürstlicher Resident' for Lower Saxony, having been faithfully nursed by his family. His place of burial, which he desired to be in Magdeburg, is not known.

ROBERT BOYLE (1627–1691)
EDMÉ MARIOTTE (1620–1684)

THE knowledge of Guericke's experiments with the air-pump and of the properties of the air which were revealed by them, soon spread over Europe, and people began to repeat the experiments in many places. One of the most enthusiastic experimenters in this field, and also in several others, was Robert Boyle. He was born in the Irish county of Cork, of an old and very well-to-do family; he made long

---

[1] See the work already referred to by Hoffmann, who is also the author of a history of Magdeburg.

journeys, particularly through Italy and France, and then
lived as a man of means at his birthplace. In his later days he
was one of the founders of the Royal Society, which became
of great importance for the development of science in
England. He was filled with a tireless passion for the observation of nature. The number of experiments of all kinds,
which he describes in his extensive works, is immense; he
was the wanderer who takes pleasure in everything there is
to be seen, and also spares no trouble to see as much as
possible, but has no great interest in plumbing hidden
depths.

He experimented with mercury in glass tubes, like
Toricelli or Pascal; he devised the plan of trapping a quantity
of air over mercury in the closed limb of a U-tube, the other
limb of which was open. Mercury columns of various
heights could then be formed in the open limb, and the
pressure of the enclosed air varied at will. He then recorded
the height of the mercury column corresponding to each
volume of air. It then appeared that, when the already
known atmospheric pressure acting on the mercury in the
open limb, was taken into account, the volume of the
enclosed air was inversely proportional to the total pressure
under which it stood. Thus the important law was discovered
which connects gas volume and gas pressure with one
another at constant temperature, and this is the most important result for posterity of Boyle's work. It was later
verified by Mariotte. Boyle himself did not attach great
importance to it.[1] As a matter of fact, when Guericke had
determined that air fills uniformly all space accessible to it,
and that the volume it occupies is the smaller, the greater the
pressure upon it, Boyle's law was practically given as the
simplest possibility. It is hardly a matter for surprise that

---

[1] In many editions of his works it is not even given, for example in
Nova experimenta de vi aeris elastica, translated from the English original
in 1659, Hagecommicum, 1661.

it was true within the accuracy obtainable at that time, and within the possible limits of pressure.

Among Boyle's numerous experiments are to be found many containing important results in chemistry, relating to the characteristic behaviour of various substances towards one another; these researches became later the foundation of chemical analysis. Examples are the detection of hydrochloric acid by precipitation with silver solution, of iron by tincture of galls, of acids by means of paper dyed with vegetable colouring matter. Boyle was altogether the first to pay close attention to chemistry simply for the sake of knowledge, and not with the object of making gold, the philosopher's stone, or an elixir of life.

Edmé Mariotte, born in Burgundy, was a priest, Prior of a monastery near Dijon; but after being made a member of the newly formed Academy in Paris as a result of his scientific activity, he lived in Paris until his death. He proved himself in many directions to be a thorough investigator. In particular, the behaviour of air at different pressures engaged his attention, in connection with the observations of Guericke, ·Toricelli, and Pascal. He also, like Boyle, arrived at the discovery of the simple quantitative relationship between pressure and volume, the 'law of Boyle or Mariotte.' His publication appeared in 1676, sixteen years later than the almost exactly similar experiments of Boyle; but Mariotte was the first to make at the same time a fundamental and important application of the law, in his essay 'Sur la nature de l'air.' He calculated quite correctly the decrease in the pressure of the atmosphere with increase in height, making use of Guericke's discovery that the density of the air must diminish as we go upwards. He therefore carried out the calculation in small steps of height. In going upwards, the density of the air is reduced as a consequence of reduced pressure, and hence equal masses of air occupy a less volume,

OTTO VON GUERICKE

OLAUS ROEMER

ROBERT BOYLE

which the law enables us to calculate. Mariotte divided the whole atmosphere – the calculus had not yet been discovered – into 4032 layers, each of which was assumed equal in pressure throughout, and calculated the height of each single layer. By summation the height corresponding to any pressure could then be calculated. This is also to-day the basis of all our knowledge of the distribution of pressure and density in the atmosphere, and in particular, of the measurement of height by means of the barometer.

Another permanent achievement of Mariotte is the determination of the actual circulation of water upon the earth. This circulation had hitherto been imagined as taking place in the main subterraneously; the springs which were seen to rise from the interior of the earth were imagined as fed by subterranean water vapour, produced by the seepage of sea-water, and perhaps also as rising from inexhaustible subterranean stores of water. The quantity of rain water was under-estimated, in so far as any attempt had been made to arrive at its amount. Even Descartes held these views. Mariotte, a scientist with Pythagoras' sense of the importance of number, based his views on extensive measurements of rainfall, which were made at his request and according to his directions, and compared the average yearly quantity, calculated as falling on the surface of the whole area of the Seine springs, with the annual quantity of water flowing down this river, which quantity he determined from the velocity of drifting ships and the area of the river bed. It appeared that the volume of rain was over six times as great as was necessary to feed the river. From this, and from numerous other observations which he made concerning the seepage and reappearance of rain water in various areas, and also concerning the variations in the water level of springs and rivers at various times of the year, he drew the well supported conclusion, that springs are fed by rain and snow, which fall on the mountains; and that therefore the circulation

Fs

of water on the earth takes place mainly on and over the surface, via the whole atmosphere. Further, Mariotte was the first to develop accurate ideas concerning the formation of raindrops in clouds, from the vapour rising up from the solid earth and from the sea.

OLAUS ROEMER
1644–1710

HIS greatest achievement was the first proof of the finite velocity of light, and an actual measurement of this velocity.

He was born at Aarhus, in Denmark, and there studied science and mathematics. From his twenty-eighth to his thirty-seventh year, he was employed at the observatory in Paris, where he was particularly interested in the movements of the moons of Jupiter discovered by Galileo. He noticed in the case of one of the moons, the inmost, that its period of revolution, measured sharply by its entry into the shadow of Jupiter, appeared to be of variable magnitude, according to the position of the earth in its orbit. When the earth was more distant from Jupiter, the moon appeared to be late; in the other half of the year, when the earth is again nearer to Jupiter, the loss of time was made good. This connection led Roemer, after concluding his observations over a period of several years, to work out an idea already familiar to Galileo, namely that light takes time for its propagation. He therefore assumed that in his observations the late arrival of the light coming from Jupiter was perceptible, since it had to follow up the earth which was in rapid motion, and he therefore calculated the velocity of light as the quotient of the semi-diameter of the earth's orbit and the total retardation in a half year. The result expressed in our present day units and based upon more accurate observation, is

300,000 kilometres a second; he laid his calculations in 1676 before the Academy in Paris. He at first met with much opposition, which, however, was for the most part only based upon Descartes' opinion that the propagation of light is instantaneous. But Huygens and Newton immediately supported Roemer; they saw in it Galileo's idea, that the velocity of light could be measured by its retardation over great distances, brought to realisation. The correctness of Roemer's conclusion was only put beyond doubt, after his death, by Bradley's discovery of aberration; and a hundred and seventy years later, the velocity of light was measured for the first time over terrestrial distances by Fizeau.

Roemer left France with many other fine spirits as a per-secuted Lutheran in the year 1681, and returned to Denmark, where he then became Director of the Observatory, and finally Burgomaster of Copenhagen. He was chiefly responsible for the general introduction of the telescope as a means of astronomical measurement; thus the meridian circle found in every observatory was his invention. His very extensive observations in Copenhagen, made with appliances of this kind, were lost, not long after his death, in the great fire which devastated the town.

CHRISTIAN HUYGENS
1629–1695

HUYGENS was Galileo's great and equally gifted successor in almost every respect; and after Kepler, the most active in preparing the way for Newton's investigations, which then for a long time represented the completion of the first insight obtained by humanity into a part of the great process of nature. He was, like Galileo, through and through an investigator of nature, who not merely struck out in certain directions and made progress of decisive importance, but took hold of everything that he met with, tested it, and strove to make it the means of discovering something new. Since possibilities of this kind were increasing at that time in number, it is not possible to discuss his work here in all its many sided aspects; but we will first state in general terms his chief contributions to the development of knowledge.

Galileo had taught us to understand the simplest phenomena of motion; starting from them, Huygens provided everything essential for carrying our understanding forward for the most complicated processes, inasmuch as he investigated certain complicated examples almost exhaustively, in particular the compound pendulum, forced motion in a circle with centrifugal force, and the phenomena of collision. He was further the founder of our insight into the wave nature of light, and the inventor and perfector of pendulum clocks; and he also investigated thoroughly important parts of geometry, particularly the properties of curves, such as the cycloid.[1]

His personal character is shown in his work, and appears also in all that is otherwise known of him; 'He shares with Galileo a noble, unsurpassable, and complete uprightness. He is quite open in stating the way by which he has been led to his discoveries, and thus introduces the reader to a full understanding of his achievement. He had also no reason to conceal this way. Though we shall see after a thousand years that he was a human being, we shall at the same time observe what manner of man he was.'[2]

The activity of the investigator of nature is compared by

[1] This is the line which a point on the circumference of a rolling wheel traces upon a plane fixed with respect to the ground line along which the wheel rolls.

[2] E. Mach, *The Science of Mechanics*, trans. by T. J. McCormack (4th Ed. 1919).

Huygens with the interpretation of a writing in cipher.[1] Everything almost at which he worked was kept back for a long time; much only appeared after his death from his posthumous papers.

Huygens was born at the Hague in Holland of a very respected, influential, and well-to-do family, which fact ensured him for his whole life independence for his work, and enabled him to travel a great deal. He studied together with his brother Constantine, who was a year older, jurisprudence and mathematics at the Hague. His first scientific achievement was mathematical, and connected with his study of Archimedes and Descartes; in it he appears as one of the founders of the calculus of probability, which has since become so important for science. His invention of the pendulum clock, the first communication concerning which he published in 1667, soon made him widely known. It was astonishing how thoroughly he developed this means of measuring time, which is to-day quite general. His work was almost exhaustive, and developed the clock in all essentials to the form at present in use; he linked with it, in the work *Horologium Oscillatorium* which appeared in 1673, a further mass of experimental results which went far beyond the actual subject.

Huygens' pendulum clock raised the measurement of time, both in science and daily life, almost suddenly to a quite new level of refinement, which had hitherto been striven for in vain, and was hardly imagined as possible. It is true that clocks with hands and trains of wheels driven by weights, had long been in existence, but they went irregularly and unreliably; for they were only regulated by frictional resistance, which is always variable. No great improvement was made when the step was taken of causing

[1] See the life written by P. Zeeman in the memorial to Huygens on the 300th anniversary of his birth (1929). There also will be found portraits of Huygens in his youth.

a somewhat massive body, rotatable about an axis, to be driven to and fro by the fastest-running wheel, in order to set a limit to the rate. Clocks of this kind were already known in Galileo's time, and yet Galileo measured time in his experiments on bodies falling on the inclined plane by the flow of water, thus using the same principle as the water clocks of the ancients, since this was found to be more trustworthy.[1] Later, when Galileo had thoroughly investigated the motion of pendulums, he himself recommended measuring time by counting the oscillations of pendulums; the difference of length of different pendulums allowing of easy calculation. Doctors began, in consequence of Galileo's proposal, to use a pendulum of thread, the length of which could be varied, to determine the pulse-rate of the sick; it was called 'Pulsilogia' (pulse measurer). What was needed for the measurement of times of any length was a pendulum which did not come to rest, and an automatic counter connected with it. Galileo already attempted this, and it was afterwards found that he had left behind suggestions for such a construction, made a few months before his death and after he had become blind; he dictated them to his son Vincenzio. The drawing is still in existence; it was carefully preserved by Vincenzio, but not made use of. Huygens was the first to actually realise the maintenance of the oscillations of a pendulum, and the counting of them, both in a very simple manner, by putting the pendulum in place of the oscillating body already mentioned and usual at the time, directly coupled to the wheelwork of the clock. The great and decisive difference between the oscillating body and the pendulum is, that the former has no fixed natural period of

---

[1] Astronomy had always used, for determining time, the movement of suitably chosen stars. Otherwise Tycho's observations would have been valueless; he only used the clocks existing in his time as secondary standards. Even the clocks of to-day must be compared with the stars from time to time, of course with a correspondingly increased degree of accuracy.

oscillation, since the force controlling it only comes from the train, and is subject to variation in friction, whereas the pendulum possesses its own controlling force in its invariable weight, and hence a fixed period of oscillation, which is uninfluenced by the power of the train; the latter being needed only to overcome the air friction of the pendulum, which is less, the heavier the pendulum is made. A new idea was brought into the old clock, which apart from the hanging pendulum was not greatly changed in form, but the effect was something entirely new: trustworthy measurement of time. Huygens says himself, in his *Horologium Oscillatorium*: 'Even if there is in the works themselves some defect or other, even if the bearings turn less easily as a result of change of temperature, as long as the movement of the clock does not cease entirely, no irregularity or retardation in motion is to be feared: the clock will either always measure the time correctly, or not measure it at all.' It is not surprising that, since the change in construction was so small, existing clocks immediately received pendulums fitted to them in place of the oscillating body. Since the year engraved on the dial, and giving the original date of construction, usually remained unaltered, the opinion later arose, that these clocks had been made as pendulum clocks before Huygens' time, but this opinion can be proved by the literature of the period to be wrong.[1]

For navigation, good clocks were and are of the greatest importance for the trustworthy determination of geographical distances, and the correct measurement of position at sea. But in this case the usual pendulum, dependent upon gravity, was not satisfactory in spite of Huygens' special efforts, on account of the oscillation of the ship. He therefore proceeded to use elastically controlled pendulums for

[1] See *Horology*, by J. E. Haswell (1928), p. 25. In England a model in the S. Kensington museum of a 'pin-wheel' escapement is usually referred to as being prior to Huygens. [Trans.]

these clocks, such as we still find to-day in the escapements of pocket watches and ships' chronometers. Here again, only the introduction of this essential feature of accurate running was the point of novelty. Pocket watches with trains of wheels, driven by a spring, existed already; they were known, on account of their external form, and their main place of origin, as 'Nuremberg eggs.' In the matter of the introduction of the spring controlled balance wheel, in place of the gravity pendulum, Hooke in England, where the need of good ships' chronometers was most pressing, had priority over Huygens, but not before the latter had realised and introduced the ordinary pendulum clock, the model for the construction of the chronometer.

Huygens further strove after extreme refinement to an astonishing degree in his pendulum clocks. His profound investigation of the peculiarities of the cycloid, which was in other respects of great value, led to his invention of the cycloidal pendulum, in order to produce constancy of period, even in the case of a large arc of swing. Though we have again abandoned the special arrangements fully described by Huygens, which are necessary for the use of this type of pendulum, it is because we have learned to work with very small arcs of swing; nevertheless, the *Horologium Oscillatorium* arouses the reader's admiration for the 'double virtuosity' with which Huygens carried out, in the most perfect manner, what had to be derived both from mechanical practice, and from the most subtle geometrical synthesis.

About the same time – from 1652 onwards – Huygens was also busy on improvements in the telescope; upon learning of Snell's law of the refraction of light, which was not yet known to Kepler and Galileo, he was able to estimate the effects of lenses with much greater accuracy than had hitherto been possible. He thus arrived also at an understanding of the errors of lenses (spherical and chromatic aberration) and their dependence upon focal length and aperture. He also ground lenses himself, assisted in part by his brother Constantine, and introduced the preliminary examination of the glass to be used as regards its freedom from striae (a process later brought to the greatest refinement some two hundred years later by Töpler in Dresden); he also improved the eye pieces of the Kepler telescope, the 'Huygens eye piece' being still in use to-day. It consists of two convex lenses, the inner of the two working together with the objective, so that the real image is formed between the two lenses of the ocular. The whole of these investigations were first published after Huygens' death in his *Dioptrics* in 1703; but he had already quite early made a remarkable application of his study of the telescope, by discovering a moon of Saturn and the free rings of this planet, an account of which he published in 1669 in his *Systema Saturnium*. Galileo had only been able to see the ring in the somewhat blurred form of an apparent multiplication of Saturn into three, and this appeared to vanish later; subsequent observers saw other inexplicable phenomena about this planet. Huygens was able, after years of observation with his better telescope, to explain everything: the ring surrounding Saturn but not connected with it is plane and thin, and has a strong inclination to the plane of the earth's orbit, but remains parallel to itself; thus it comes about that twice during a revolution of Saturn in its orbit (that is, about every fifteen years) it is illuminated on the narrow edge only, and thus becomes quite or almost invisible, whereas in the intermediate period it arrives at full visibility. Huygens made these discoveries with a telescope constructed from lenses he had ground himself; its objective, which still exists in Utrecht, was only about five and a half centimetres in diameter, but had a focal length of three metres; the ocular was a small lens of seven centimetres focal length.[1] After the telescope, Huygens also worked on

[1] See *Oeuvres complètes de Christian Huygens*. The only English translation of Huygens is of the *Treatise on Light* (S. P. Thompson, 1912).

the microscope, and introduced dark field illumination,[1] which, when improved by modern means, gave us the ultra-microscope.

These discoveries, together with the invention of the pendulum clock, resulted in Huygens being called to the newly-founded Academy in Paris, where he occupied an influential position with a substantial salary and free quarters, from 1666 to 1681. Thereafter, Louis XIV went to war with Holland; the minister Holbert, who had a high appreciation of Huygens, lost his influence, and the banishment of the Protestants (Revocation of the Edict of Nantes, 1685) threatened. Huygens therefore gave up for lost all that France had offered him, and returned to Holland to his paternal estate Hofwijck near the Hague, where he died at the age of sixty-seven, after fourteen years of further uninterrupted activity. He did not found a family, and otherwise remained a lonely man. Only a journey to England occurred in this latter period of his life, where he made the personal acquaintance of Newton, his fourteen-year younger contemporary. Newton had a high opinion of him, and would have liked to keep him at the University of Cambridge, but nothing came of this.

Huygens' consideration of the movement of the pendulum first led him to discover the present well-known formula for calculating the time of oscillation from the length, which formula summarises Galileo's laws of the pendulum, and further gives the exact connection with the acceleration of gravity which is operative in free falling.[2] He also saw that constancy of the period of oscillation with varying arc or amplitude cannot hold good exactly for the ordinary pendulum, but only for the cycloidal pendulum, in which the bob moves along a cycloid, instead of along the ordinary circular path. He arrived at this view by recognising

---

[1] See *Oeuvres complètes*, vol. 13, 2, p. 696 (1692).
[2] In *Horologium Oscillatorium*, second part, paragraph 25.

as the essential point in the movement of the simple pendulum the predetermined path of the bob, along which path the bob moves under the influence of gravity, according to the laws already given by Galileo for motion on an inclined plane; the latter had also himself investigated falling on a circular path.

Further consideration of the motion of the compound pendulum, that is, an oscillating body which cannot even approximately be regarded as a single point (as for example a swinging rod), led Huygens deep into a knowledge of the essential features of rotational motion generally. This opened the way to a general mastery of every kind of motion of solid bodies, including also the heavenly bodies: the problem of linear motion under the influence of given forces having been solved in all essential respects by Galileo. If we attempt to follow the main lines of Huygens' thinking in this important matter, we realise that even advances in knowledge relating to very complicated matters are based on simple trains of thought, though the complete working out may be very elaborate.

When a rod hung up at one end swings as a pendulum, each of its single parts swinging about the given axis forms a simple pendulum, and the periods of oscillation of all these single pendulums can be calculated from their lengths. The shorter of these will swing faster, and the longer ones correspondingly slower. But all these single pendulums are rigidly connected together to form a rod, and hence are obliged to move at the same rate, and the question is, what this rate will be. Huygens arrived at a correct answer by making use, as his great countryman Stevin had previously done, of a generally known and therefore well assured existing experience. It related to the centre of gravity. In the present case this tells us that when the pendulum rod moves under the influence of gravity, nothing will ever happen to raise its centre of gravity to a greater height than that from which it originally descended, even when the pendulum during its movement has been subjected to effects which have nothing to do with its weight. Galileo had already convinced himself of this in the case of the simple pendulum, by letting such a pendulum swing both freely and also past a projecting peg, which caught and bent the thread in the course of the movement, so that the bob of the pendulum was forced into a new circular path with the peg as centre, from the moment when the thread met the latter. In such experiments the bob never reached a greater height than that from which it had been released, and when it attained a less height, this was obviously only the result of frictional resistances which had nothing to do with the main process. Huygens worked out in an exactly similar manner, by means of an imaginary experiment, the case of the oscillating rod. Imaginary experiments became, after Stevin's time, of ever increasing importance in research, but they must be of a permissible description; that is to say, they must only deal with processes that could be realised with sufficient approximation. We imagine the rod to be let loose and to swing until it reaches its lowest or vertical position, and at the moment when it reaches this, we suppose that the connection between its parts is abolished, so that these now continue to swing as simple pendulums. Even then, the combined centre of gravity of all the parts of the pendulum can only rise to the same point from which it had descended. But the motions of all the parts are known, since they are simple pendulums. These pendulums were released with the velocities which they had at the moment of separation, that is with velocities which were proportional to their distances from the axis of rotation. Hence if one of these velocities is known, they are all known, and the motion of the whole centre of gravity while they swing further as simple pendulums, is also known. The question now to be settled by simple and known processes of calculation is conversely the following: how great must have been the velocities of the paths at the moment of separation, and hence also that of any given point of the undivided pendulum, in order that the total centre of gravity shall only reach the definite height. When this velocity is calculated, the height from which the oscillation began being given, we have the desired period of oscillation of the compound pendulum, or also the length of the simple pendulum having the same time of swing.

When this calculation is made, a sum appears at an important point, formed of all the products of each mass particle of the compound pendulum and the square of its distance from the axis of rotation. This sum, first grasped by Huygens, gives the key to the calculation of all rotational motion of every kind; it later received the name of the 'moment of inertia' of the body in question about the given axis. The moment of inertia is the measure of the inertia for all rotational motion, just as is the simple mass for all linear motion. The part played in this case by the squares of the distances of the single mass particles from the axis depends upon the fact, as we can see from Huygens' argument, that according to Galileo's laws of fall, the square of the velocity of a body is a proportionate measure of the height to which it is able to rise along any path – the choice of which makes no difference – against its own weight, and that in the case of rotary motion, the velocities are proportional to the distances from the axis.

Hence we find in the work of Huygens for the first time the product of the mass and the square of the velocity appearing as a quantity of importance for phenomena of motion. Particularly in investigating the process of the collision of elastic bodies, the importance of this product is made clear by Huygens, inasmuch as he shows that the product in question, when summed up for all bodies entering into collision, is not altered by the latter, however the velocities may change. It is true that Huygens generally uses the term weight, and often even size, instead of mass, in reference to the body investigated. The peculiar nature of the concept of mass was only cleared up by Newton in the further prosecution of these investigations.

Research into the processes of collision presented quite peculiar difficulties, though the processes are common enough in everyday life; their peculiarity is that they take place in very short periods of time. Galileo himself already paid attention to collision, and contemporaneously with Huygens (1688), the Englishmen Wallis and Wren (the famous architect),[1] Wallis investigated the collision of inelastic bodies, Wren and Huygens that of elastic bodies, and the essential point is that they found simple relations between the magnitude of the masses concerned in the collision and their velocities before and after it, without however needing to investigate the details of the process taking place in the very short period of contact of the two bodies. Huygens in this respect was far in advance of the others, and the manner in which he handles the matter gives us remarkable evidence of his gifts as a scientist. In a manner quite new at the time, he connects the discovery already made by Galileo, that velocities of different origin present together do not interfere with one another, with the fact, already used by him in discussing the compound pendulum, that the centre of gravity does not increase in height. From these premises, with the addition of a few experimental facts derived from simple cases of collision, he deduced a large number of the most important laws concerning collision. The agreement of the laws thus found with the experiments carried out upon 'hard' elastic balls, welded the knowledge of the phenomena of motion into a consistent whole.

In all these investigations of Huygens, we meet for the

[1] Not long afterwards we have Markus Marci in Prague (see Mach, *The Science of Mechanics*, trans. by T. J. McCormack (4th Ed., 1919). Mariotte's essay on collision, which is more often referred to, came much later (1687).

first time the *dynamic* importance of the centre oᵢ gravity, and not only its original, *static* bearing, known since the time of Archimedes. A further discovery is the importance of the product of mass into velocity, to which the name of quantity of motion, or momentum, was given.[1]

A further fundamental step, especially in view of Newton's work which soon followed, was taken by Huygens in determining the laws of centrifugal force, which appears in rotary motion, and in fact whenever bodies of any kind move in curved paths. He was not only the first to observe it, but also to investigate it thoroughly. The law which he found, according to which this force, which is always directed away from the centre of curvature of the path, is proportional to the square of the speed in the path, inversely proportional to the radius of the path, and proportional to the moving mass, was first published by him in his essay on the pendulum clock (1673); but the working out only appeared, together with that of the laws of collision, after his death from his posthumous papers. But the deduction is based on nothing more than Galileo's law of inertia, and this variety of force, as a pure consequence of inertia, was so thoroughly investigated by Huygens, together with its effects, that even to-day there is nothing of importance to be added.

We already find in the work of Huygens on phenomena of motion the elements out of which Newton was soon able to build the whole structure of mechanics, including that of the heavenly bodies.

Very different in nature from these investigations of motion are Huygens' investigations of the nature of light. In

---

[1] Huygens already, in 1669 (in the *Journal des Savants*) calls the tendency to constancy of this product, summed up over a whole system of bodies, 'une admirable loi de la nature.' Descartes already made use of the same product in calculation, but he did not yet realise that its all-round importance is bound up with a consideration not only of the magnitude, but also of the direction of velocity (treating it as a vector quantity, in other words), and therefore, for example, introducing velocities in opposite directions with opposite signs into the calculation.

this case, he himself supplies an entirely new foundation based on observation, and supported by entirely new arguments. What he found as known, were the rectilinear propagation and reflection of light, understood since Euclid's time; the refraction of light, studied by Kepler and, more accurately by Snell, together with its successful application to the telescope by Galileo; and finally, and of particular importance, the propagation from the source to the eye with a limited and known velocity, determined by Roemer only a short time previously. Huygens combined these and all other known facts concerning light and its production, and concluded that we must suppose it to consist of a vibration which spreads outwards from the source, just like sound in air. The medium, however, in which the light is propagated as vibration, cannot be the air. Huygens deduces this from Guericke's, and later Boyle's observation of the propagation of light through a space empty of air, and then draws the further conclusion that space free from all matter, even air, must still contain a something capable of permitting vibration to be transmitted. This something is called by Huygens the *ether*. With him, therefore, we have the beginning of the physics of the ether. The chief point, however, is the proof that light actually, and in every respect, possesses the peculiar properties of a state of vibration which is transmitted in waves, and here Huygens contributed the first pieces of evidence which have been of importance ever since. He first of all shows how the property of light waves, coming from any direction, of penetrating one another without hindrance or disturbance, is precisely what we should expect from the propagation of vibrations; it is found also in the case of sound, and, as he showed, in the collision of elastic bodies. Here he makes the observation, founded on experiment, that balls in collision experience a momentary flattening, a process obviously requiring time, so that it is clear that a blow transmitted through a row of

CHRISTIAN HUYGENS

ISAAC NEWTON

balls must experience a certain retardation, in exact correspondence with the finite velocity of light. In sources of light, for example a candle flame, he assumes that each small luminous particle produces waves independently, but that all the single waves unite to form larger wave fronts, which then combine and move forward as the beam of light – an idea entirely in accordance with our present day view.

Concerning the propagation of the wave-front, Huygens then states a law, which has hitherto always been found to hold for wave propagation, and is described as 'Huygens' principle.' According to this, every part of the medium of propagation which is disturbed by a wave, acts for its part as a new centre of propagation for its own neighbourhood, whereby the elementary waves proceeding from a single point of the medium, again join together to form new wave fronts along their common surface of contact. By means of this principle, Huygens was able to explain completely the reflection and refraction of light, and the known laws relating to these. In the case of refraction, we must assume that the velocity of light in the refracting medium, for example, glass or water, is smaller than in free ether, by an amount proportional to the index of refraction. This was shown by Fresnel a hundred and forty years later to be true, and forty years later still, a further and more direct demonstration was given by Foucault.

Huygens himself carried out another and special test of these ideas, by applying them to the highly complicated phenomenon of double refraction, which was known in the case of Iceland Spar, but had remained completely incomprehensible. In the case of one of the two rays which result from a single original ray upon refraction in the Spar, the one called the 'extraordinary ray,' no law of refraction exists; according to the circumstances it takes very various directions, by no means to be expected from Snell's law of refraction, which holds for the other, or 'ordinary,' ray.

Gs

Thus for example, refraction may take place upon normal incidence. Huygens was able to make all these peculiar phenomena intelligible and predictable by means of his principle, by the simple assumption that in the crystal, a part of the light is not propagated in all directions with the same velocity, but more slowly in the direction of the crystalline axis, and with increasing velocity in directions at an angle to the axis; fastest of all in a direction at right angles to it. The supposition that light may travel at different velocities in different directions through the crystal was not improbable, since in other respects, for example as regards their mechanical strength, crystals generally exhibit the peculiarity of possessing different properties in different directions; this assumption also proved to be of great value. Huygens compared the conclusions drawn from his principle of the propagation of light with reality by means of numerous observations on Iceland Spar, and in part also on quartz (rock crystal); in this connection, he also made his important discovery of the polarisation of light. He recognised that rays of light in passing through a first crystal of Iceland Spar are given 'a certain form or arrangement, according to which they behave differently, according to the position in which they meet a second crystal of Iceland Spar.' The nature of this certain form was not determined until a hundred and forty-three years later, by Fresnel. We see how far Huygens had progressed in these investigations. They were first published in the year 1676 by a discourse to the Academy in Paris, and then in 1690 under the title *Essay on Light*, one of the most remarkable documents of early great progress in scientific investigation. The last essay published by Huygens himself, called *Cosmotheoros* and dedicated to his brother Constantine, discusses thoroughly the idea previously put forward by Giordano Bruno regarding the multiplicity of planets similar to the earth in space, fit for the habitation of living beings and probably also inhabited. Huygens enjoyed, in respect of this idea also, the advantages of a country in which, thanks to Luther's deeds, he had not to fear the anger of the Pope. Newton, who shared this advantage with him, showed by his success even more clearly its importance for the progress of science.

ISAAC NEWTON
1643–1727

NOT quite a year after Galileo's death, a weakly child was born in the village of Woolsthorpe in East Anglia, whom it was hardly hoped to rear, but who was destined to be the author of one of the greatest revelations which mankind had ever received from scientific investigation. Newton discovered universal gravitation, the force which holds the heavenly bodies together, guides earth, planets, and moon in their paths, acting according to a fixed and highly simple law; while it is also the force which causes a stone to fall to the ground, or to describe its trajectory. Indeed, it acts between all matter, earthly or heavenly, small or large. For the first time therefore, the whole of the visible world appeared as a single great unity; heaven and earth ceased to be opposed to one another. Nevertheless, the Universe only became more astonishing than ever before, when everything was found to form a single whole, united by an all pervading and uniformly acting force, to which all visible motion is subjected, from our nearest surroundings on earth to the whole solar system and likewise to the other more distant suns, which revolve around one another. A new insight was given to man, which once grasped, should have raised him spiritually, and made him better and nobler. This great increase of our knowledge concerning the nature of the world in which, at first completely ignorant, we find ourselves, was brought forward and supported by Newton in the work *Philosophiae Naturalis Principia Mathematica* ('The Mathematical Principles of Natural Knowledge' would be a suitable translation). At the same time he points out the way to all future investigation, and indeed, as regards mechanics – the phenomena of motion of massive objects – he even completes the investigation in all fundamental respects. If we open the book in order to examine it in detail, we are astonished, quite apart from the main discovery, by every part of it, and overwhelmed with admiration for the greatness, the extent, the power, as well as the fineness of structure, of what he erects upon the foundations given by Pythagoras, Archimedes, Leonardo, Stevin, and in particular Galileo and Huygens, using the material afforded by Copernicus, Tycho and Kepler, and the tools provided by mathematicians from Euclid to Descartes, with very essential additions of his own. We are not less astonished, and almost overwhelmed, by the countless number of single achievements, which, from whatever side the work is regarded and studied, are revealed to anyone capable of comprehending them. The whole, when we consider the richness of its contents, the general and detailed form, and the impression it makes of towering greatness, can only be compared to a grand old Gothic cathedral: one stands in front of it filled with astonishment, absorbed in gazing at it, and without words to express one's impressions. Great cathedrals were built in numbers by the masters of Gothic; but among the works of men of science, Newton's *Principia* is unique of its kind. The artist is in a different position from that of the man of science; he can create and work from his own mind without limits, and in so far as he possesses the necessary materials, he will always produce work corresponding to the greatness of his powers. But the scientific investigator depends in his work, apart from the favourableness or otherwise of external circumstances, upon the laborious and exhaustive collection of natural knowledge, which can only be obtained from outside, from the world as it is, – and it is generally surprising and peculiar in an unexpected way. Without a great store of knowledge thus collected, no great work of science can result, and Newton was fortunate in being in possession of the work of so many rare spirits which we have just reviewed, from Archimedes to Huygens. But the manner in which he grasped this knowledge, and greatly increased it on his own account, was worthy of the great achievement of his predecessors, and forms a monument to them and to him; science should be thankful to possess such a monument.

Coming now to consider the new matter which he himself added, we are compelled by its richness to limit ourselves somewhat – even more than in the case of Huygens – but not to the point of being unable to give a fairly complete picture.

One of Newton's especial achievements was his elucidation of the fundamental concepts: mass (*massa* or *corpus*),[1] weight (*pondus*), force (*vis*). Without this elucidation, the *Principia* could not have been written. Mass means for him – and for us to-day – always the measure of inertia, that is, the quantitative expression of the tendency, discovered by Galileo, of every body to oppose change of velocity.[2]

[1] Newton wrote his *Principia* in Latin; his *Opticks* was written in English.

[2] This meaning of the word 'mass' in Newton's work entirely corresponds with its application throughout the *Principia*. The fact that Newton did not expressly define the concept of mass in this way, has often been used as a reproach to him. But we must remember that concepts in science do not by any means rest upon the definition of words, but that, on the contrary, a concept must gradually be formed hand in hand with the growth of experience, in order that it may be of such a nature as best to serve the representation of observed reality. When growing experience has brought us to such useful concepts, we may then proceed to describe in words this mental possession, that is to say, to put forward a definition of the concept. What critics may therefore rightly think is the following: that Newton arrived at the complete possession of the concept of mass as we have it to-day, but that he left posterity to produce an improved verbal definition of it, and also to test the usefulness of the concept further. The addition which experience has led us to make to Newton's concept of mass is of quite recent date; it belongs to the last of our series of men of science. It has caused no change in the concept itself.

Force is every cause of change of velocity. Weight is the force of gravity, which acts upon the body in question on the earth. The fact that mass and weight are proportional to one another for all bodies, is clearly stated as an experimental fact.[1] The fundamental experimental facts had already been recognised by Stevin and Galileo, in the equal rate of fall of all bodies, and in a more refined form by Galileo in his discovery of the equal rate of oscillation of all pendulums of equal length.[2] The latter fact was tested by Newton with still greater accuracy than Galileo, using pendulums of material of such different specific gravity and nature as gold, silver, lead, glass, sand, rock salt, wood, water, wheat, and he found it to be confirmed in every case.[3] In these experiments he took into account the frictional resistance experienced by pendulums, in common with all moving bodies, at least in the air, whereby he brings by way of preface a comprehensive investigation, which itself has become of fundamental importance, concerning frictional forces in various media.

A second great advance of Newton's is the clear statement of three laws of motion for all matter. The first two are: (1) Galileo's law of inertia and (2) Galileo's law of the proportionality between force and acceleration, whereby however Newton, instead of taking acceleration, that is the change of velocity in unit time, takes change of momentum, that is to say, the change in unit time of the product of mass and velocity, which had appeared since Huygens' investigation as a measure of the effect of forces. It is clear that this refinement, introduced by Newton as compared with Galileo, is only of importance as regards variable masses (and also moments of inertia). But it is also true in this case even for non-material masses, as has been shown by the very latest experimental results.[4] It is directly clear that the law of

---

[1] *Principia lib.* 1, *def.* 1. [2] *Dialogues*, first day. [3] *Principia lib.* 3, *prop.* 6.
[4] See my statement in *Über Aether und Uraether*, 2nd Ed. (Leipzig, 1922), p. 48.

inertia is only a special case of the general second law of motion, but one which, on account of its great importance, appears to merit a special statement. Force, according to the definition of the second law, is decisive for all change of motion. But here we have the peculiar fact that force can be detected and measured without change of motion, namely in cases, investigated by Archimedes, of equilibrium between forces; for example the force of gravity, or weight, in the case of a balance. This is taken into account by Newton in a special analysis of the concept of force.

The third law of motion is one introduced by himself; it is the law of the equality between action and reaction. It deals with a peculiarity of all forces in nature, namely their occurrence in pairs, so that to every force acting in any direction, there belongs another force acting in the same straight line, and of the same magnitude, but opposite in direction. Newton derives this law expressly from experience – very simple and everyday experience, of which he gives examples – just as Galileo derived the two first laws from experience. Indeed, every person of sense must regard it as self-evident, that everything relating to natural science can only be derived from experience. To experience then belong all observations made to test the correctness of conclusions drawn from laws which, perhaps, were originally only founded upon a small degree of experience.

These three laws of motion occupy a further peculiar position in Newton's case; they are given by him a meaning hitherto unthought of, even by Huygens, namely that they are simple and self-evidently valid for all matter, including that in cosmic space. This makes Newton the founder of the mechanics of the heavens. The fact that he could devise this mechanics successfully – and he did so to a point which has left us nothing of importance to add up to to-day – using laws of motion derived only from terrestrial matter, the fact in other words, that there are laws of motion valid for all matter, could certainly not have been foreseen; Newton had rather to provide the proof by successfully developing his mechanics of the heavens, and this meant, even without the simultaneous discovery of the general law of gravitation, in itself a great and new insight into the nature of the world. It may be remarked by the way, that the universal validity of the laws of motion is better assured than the action of gravitation over space of any extent. We know from the motion of the farthest double stars, which can be determined by Doppler's principle, that the phenomena of motion there taking place are of the same kind as, and take place under forces similar to, those in our solar system, but we have no definite assurance that the gravitational forces between our sun and stars at any distance act according to the square of the distance; the case might also be otherwise.

Newton's three laws of motion are exhaustive even to-day for all known phenomena of the motion of matter, and in part even further; and in so far, dynamics (the doctrine of the motion of matter) reached with Newton its full and fundamental development.

By the strict application of the laws of motion to the heavenly bodies, Newton arrived at a knowledge of the forces which act between them. The fact that such forces must exist, say between earth and moon, was shown by the curvature of the moon's path; without the action of force, it would only possess uniform motion in a straight line, according to the law of inertia. The case of the moon is the same as that of a body projected horizontally near the earth; this body likewise describes a path curved in the direction of the force acting upon it. If the velocity of the body projected horizontally could be made sufficiently great, it would no longer fall to the ground, but travel around the earth like the moon, which is drawn towards the centre of the earth by the effect of gravity.[1] Forces must also be present in the case of all

---

[1] *Principia, lib.* 1, discussion of definition 5.

the known paths of the moons and planets, which are directed from all sides in each case towards a fixed point. Newton thus arrives at a thorough discussion of such 'centripetal' forces, concerning which he states a large number of theorems. He here covers in part the same ground as Huygens with his 'centrifugal' force;[1] for centripetal and centrifugal forces are equal to one another, but opposite in direction, in the case of a circular path. Newton then investigated the form of the path, and the motion in it, in the case of various laws of centripetal force, both with and without frictional force, which latter may again be proportional to the first or second power of the velocity; here a great development of geometry took place.[2] It appeared that motion according to all three laws of Kepler, only occurs when the centripetal force is directed towards a focus of the elliptical path (more generally a path in the form of a conic section), and when it acts according to the inverse square of the distance, and is proportional to the mass, all frictional forces being absent. According to the third law of motion, the force between sun and planet, earth and moon must be mutual, and it follows from this, that both must be in motion about their common centre of gravity, and also that the masses of both enter equally into the magnitude of the force. The law of the actual force acting between sun and planets, gravitation, was thus grasped. Since the four moons of Jupiter and the moons of Saturn (in Newton's time five had already been found and observed) act according to the same laws, as Newton shows in detail, the law of gravitation is also proved for these bodies, and finally also for the comets, since, as Newton, with the important assistance of his pupil Halley, shows in the case of several of them, they move in conic sections with the sun as

[1] Newton recognises Huygens' previous work by a special note (*Principia, lib.* 1, *sec.* 2, *prop.* 4, *scolium*).

[2] Newton also studied for the first time, in a special essay, lines of the third order; altogether, he made very great extensions of the analytical geometry founded by Descartes.

focus. The law of gravitation is thus shown to hold for any two masses of the planetary system; they attract one another with forces proportional to the two masses, and inversely proportional to the square of the distance between them.

Newton already regards as a conclusion of this the existence of disturbances, exerted by the planets upon one another through their mutual forces, and thus going beyond the laws of Kepler. He considers in particular the influence of the gravitational force between the sun and the earth's moon upon the path of the latter, and is able to give a full explanation of the irregularities in the motion of the moon. If the law of gravitation is true for any two masses whatever, it must also be true for any two parts of the earth; and therefore for a body on the earth's surface, and the whole mass of the earth. The resulting force upon the body, which is known as its weight, must be the resultant of all the single forces acting between the given body and all single parts of the earth, in accordance with their masses and the square of their distance. This brings Newton to calculate the resulting forces for spherical masses such as the earth, and partly also for non-spherical. He finds that the resulting force for any point outside the sphere has exactly the magnitude and direction that it would have if the whole mass of the sphere were concentrated at its centre. The attractive force inside the sphere is given by another law; here the force decreases as we approach the centre of the sphere, in simple proportion to the distance from the centre; from this we know the force of gravity in the interior of the earth. Newton thus arrives at a correct comparison between gravity on the earth's surface and that acting upon the moon, the latter being calculated from the centrifugal force of the moon, and hence from the distance and period of rotation of the latter. It is found, that making use of the size of the earth's radius, first determined in Newton's time with sufficient accuracy, that in

actual fact the force of gravity falls off from the earth's surface to the moon, in accordance with the square of the distance, and hence that the force of gravity known from the
weight of objects upon earth is actually only a special case of
universal gravitation.[1]

Newton also calculated the flattening of the earth at the
poles resulting from centrifugal force, at a time when the
opinion was still held by some that the earth is an ellipsoid
lengthened in the direction of the axis. He also calculated
the increase in the force of gravity on the earth's surface as
we pass from the equator to the poles, and thus explained
hitherto inexplicable observations concerning the change in
the rate of pendulum clocks when brought into other lati-
tudes; these observations could then be used to investigate
the form of the earth. It now became for the first time
directly obvious that, while weight and mass are exactly
proportional for all bodies at the same place, the weight of
any body can alter independently of its mass, a fact which
follows from the law of gravitation.

An especially valuable achievement was the discovery of
the true and complete explanation of the tides of the sea,
which had been known from the earliest times. Galileo and
Kepler had striven after an explanation in vain, although
the latter already assumed an attraction exerted by the moon
upon the water of the earth, just as he also considered from
time to time an attractive force proceeding from the sun and
acting on the planets. The attractive force of the moon
could only explain the heaping up of the water on the side
turned towards it, but not the same effect, always simultaneously present, upon the side turned away from the moon.
All these phenomena result, as Newton showed, from the
simultaneous action of gravitation and centrifugal force,
the latter being produced by the fact that the earth also
revolves once a month, namely together with the moon about

[1] *Principia, lib.* 3, *prop.* 4.

the common centre of gravity of earth and moon. Furthermore, the effect of the sun has also to be taken into account,
which Newton also did. But he goes further, inasmuch
as he did not merely give a correct and complete qualitative
explanation of the tides, but carried out calculations which
enabled him to find from the known heights of the tide the
actual relationship between the mass of the moon and that
of the earth.

It was thus possible not merely to ascribe masses to the
heavenly bodies according to a well-defined concept, which
had never previously been imagined, but even to determine
the magnitude of these masses. The calculation of the mass
always takes place according to the law of gravitation; it is
only possible when the body in question produced sensible
motion in another body, as when the moon acts on the earth's
water. The earth, Jupiter, and Saturn, move their moons;
thus Newton was able to calculate the mass of these planets.
The sun moves all the planets, and he is thus also able to
deduce its mass. Knowing the mass of all these heavenly
bodies, sun, planets, and moon, he is accordingly able, since
their diameters are known, to calculate the specific gravity
of the material of which they are made, in which calculation
the earth always serves as the standard of comparison. He
finds for example that the moon consists of specifically
heavier material, Jupiter and the sun of lighter material,
than the earth. By then estimating the mean specific
gravity of the earth quite correctly as between 5 and 6,[1]
he is able to give the specific gravity of the other heavenly
bodies in ordinary measure. He was even able to state the
strength of gravity on the surface of the moon, Jupiter and
the sun. What a difference as compared with past
times in our knowledge of the unapproachable heavenly
bodies! Through Newton we almost feel at home upon
them.

[1] *Principia, lib.* 3, *prop.* 10.

Newton also calculated the motions of the earth's axis, and the consequences of them: viz., the precession of the equinoxes, and the nutation superimposed upon this.

A very remarkable conclusion was that the whole centre of mass of the solar system was either at rest or in uniform motion in a straight line. This is a great example of the centre of mass principle, in part recognised by Huygens, and fully set forth by Newton in his third law, according to which the centre of mass of a system of bodies, or its momentum, remains uninfluenced, when all forces and their reactions only act within the system. Only when forces act upon the system from without, forces whose reactions operate outside the system, does the centre of mass experience an acceleration in accordance with the second law, and the momentum undergo a corresponding change.

These extensive investigations were rendered possible for Newton by his discovery of a new method of calculation, that of 'fluxions,' as he calls it; it was later called by Leibniz the 'infinitesimal' or 'differential and integral' calculus. This is a method of reckoning with infinitely small quantities, the need for which appeared very early. Galileo was already obliged, in following out the simple motion of free fall, to divide up the whole time of fall into parts, in order to be able to consider the velocity in each part of the time singly, and similar problems arise in the investigation of every kind of motion. But since the velocity changes uniformly, and not in jumps, strictly accurate calculation requires the single periods of time to be infinitely small, whereby the distances corresponding to them also become infinitely small, and the question remains of how we are to regard the relation between distance and time, the velocity, as a relationship between two infinitely small quantities, and how we are to calculate it. In the same way in Descartes' calculation of the rainbow, and Mariotte's calculation of the distribution of pressure in the earth's atmosphere, it was necessary to undertake a similar division and execute a large number of single calculations, and this could only really be done satisfactorily to any desired degree of accuracy, if the division had been carried to an infinite number of infinitely small parts. Newton recognised that as we proceed to a finer and finer division, the relationship between two corresponding magnitudes, such as distance and time, approaches a fixed limit, although the magnitudes themselves, taken singly, continually become smaller. The essential point of these problems is always of this nature. His special mathematical studies, which we cannot discuss here, but which in themselves were fruitful in results, enabled him to propound rules for the direct calculation of such limiting values, in the case when the division becomes infinitely small. This is the fundamental idea of Newton's calculus of fluxions, which gave him a correct solution of many mathematical problems hitherto only soluble with great difficulty, or not at all.[1]

The *Principia* is not by any means confined to the mechanics of the heavens. It develops and increases our whole knowledge of nature quite generally, and in a fashion fundamental for a long time subsequently, so that many later achievements in single departments appear unimportant by comparison, since they did not bring anything essentially new, but only refined upon what was already known. In particular, the fundamental lines of the phenomena of motion in liquids and gases (hydrodynamics and aerodynamics), in which department only Toricelli's law existed, were developed almost exhaustively by Newton as regards characteristic single cases.[2] He investigates phenomena of flow and formation of eddies in liquids, oscillations of liquids

[1] The method used by Newton for this purpose is not given in detail in the *Principia*. Since our present-day, easily learned, method of calculation was only later fully developed by Leibniz, what Newton managed to achieve at that time is the more astonishing.

[2] *Principia*, book 2.

in U-tubes, propagation of waves, internal friction or viscosity, the resistance to projectiles, damped oscillations, and many similar matters. He thus became the founder of the science of sound (acoustics). He developed, as the result of a profound insight into the phenomena of waves, the method of calculating the velocity of sound, finding it to be equal to the square root of the ratio between elastic force (pressure of the air) and density (specific gravity), and concludes from this that the velocity of sound must increase with increase of temperature. The calculation agreed with that already found since Leonardo's time by observation of echoes. It was a hundred and thirty years later that something new was added to this by Laplace, whereby agreement with later and more refined measurements of the velocity of sound was arrived at. Newton also developed the fundamental and simple relationships, holding for all wave processes, between velocity of propagation, wave-length, and period of vibration, or frequency. He then calculated the wave-length of a note of known frequency, and found it to be double the length of the open pipe producing the note.[1]

The velocity of water waves, and its dependence upon the wave-length, was also calculated by Newton directly (for small amplitudes), with the assumption of rectilinear motion of the particles of water; but he also remarks that the motion really takes place in circles, which fact was further developed

[1] As regards the determination of the frequency, Newton refers to Sauveur in Paris, who was the first to determine absolute frequencies of musical notes. This was done by counting the beats – which Sauveur also interpreted for the first time correctly – between two deep notes, which were related to one another as fundamental and second, whereupon the two unknown frequencies could be calculated from the difference and the ratio. All other sounds could then be given in absolute terms by ear, making use of Galileo's proof of the fixed relationship between the frequencies of the musical intervals. Sauveur (1653–1716) was also the first to investigate thoroughly the different forms of oscillation of stretched strings, which are responsible for the harmonic overtones. For this purpose he used paper riders, and gentle contact, in the same manner as to-day, and he invented the terms 'loop,' 'node,' and 'stationary wave.'

by Wilhelm Weber a hundred and forty years later. Hydro-dynamics, developed further upon Newton's foundation, has also succeeded in calculating the velocity of propagation for any amplitude, but nothing of fundamental novelty concern-ing these gravitational waves has been added.

The wide scope of his *Principia* is already stated by Newton himself in his mention of those parts of science which were then only present in their beginnings, or where merely indicated as possible, and of which he says that we have not sufficient experience (*Copia experimentorum*), to be able to say anything of a definite nature.[1] He mentions the forces with which neighbouring particles of a body attract one another over the shortest distance, so as to hold the body together; from which it appears that he did not regard these forces – the molecular and the chemical forces in our present-day terminology – as did Galileo and even Huygens, as due to external pressure, but as acting like gravitation, but, being different from it. He also mentions electrical attrac-tion and repulsion acting over greater distances, light and its phenomena and heating effects, and the phenomenon of life. The latter phenomenon is one which we, two hundred years later, still regard in all its aspects, with an amazement, and practically complete ignorance, as great as Newton's.

This is true to almost the same extent concerning the cause of gravitation (*causum gravitationis nondum assignavi*).[2] We only know as Newton did, that no matter is without it, that it proceeds from every single particle of the whole volume of every body, acts through everything, and main-tains the law of inverse squares to the greatest distances; and we are obliged – if we leave aside assumptions which are not yet sufficiently tested – to continue to say with Newton: 'Enough that the existence of gravitation has become clear,

[1] Conclusion of the *Scolium generale* at the end of the *Principia*.
[2] *Scolium generale*.

and that its action according to the laws put forward by us is proved, and sufficiently explains the motions of the heavenly bodies and of our sea.'[1]   While in the last two hundred years we have made some progress in the physics of the ether, to which other phenomena, rendered more comprehensible since Newton's time, belong, we can only suppose that gravitation also may be a phenomenon connected with the ether. The phenomena of life are seen to have their roots beyond the region known to us as the ether, just as the ether lies beyond matter as regards its accessibility to our understanding, but not without being very closely connected with matter; a sign that everything in nature, both nearest and farthest, is everywhere interlinked.

'Hypotheses non fingo' says Newton[2] in face of the limits of his knowledge; and he means by this: 'mere supposition – what has not been properly deduced from phenomena – is not put forward by me as science.'   The warning therein contained is certainly of permanent validity, if the investigation of nature is to remain what all great men of science have held it to be: the collection of knowledge of the truth.

Newton regards the question of the existence of God as not outside the range of science; for he treats of it shortly in the *Principia* at the close of the book.[3]

---

[1] *Scolium generale.*
[2] *Scolium generale.*
[3] End of the *Scolium generale.*   We may note particularly the following amongst all that Newton says concerning the Divinity: 'Totus est sui similis . . . ; sed more minime humano . . . more nobis prorsus incognito' (Wholly like to itself, but in no way human in nature . . . of a nature entirely unknown to us).   This should most of all prevent erroneous interpretations, which certain other of Newton's remarks have caused those to fall into, who are not able to distinguish between words coming from Newton's own mind, which was instructed in so high a degree by the study of nature, and words which he took from the study of the Old and New Testaments, the interpretation of which by the theologians he obviously trusted much more than can the scientist of to-day, when historical knowledge is much further advanced as regards the composition of these writings.

He is right in so far as he sees in all completely understood knowledge the direct action of the Divinity,[1] and inasmuch as a Divinity not recognisable in all natural events, also those in the domain of the non-living, would in no way deserve the name, which is linked with the highest endeavour of humanity, reaching out beyond its own limits.   More or less understood knowledge thus appears, from Newton's point of view, as Divine action become more or less intelligible in some of its details.   None of the greatest men of science has ever maintained that when it becomes partially intelligible it is any less wonderful than when entirely unintelligible: as are living organisms, for example.   All their statements testify to the opposite.   The results of investigation show this directly; every new piece of insight obtained, as that into the Keplerian laws by means of gravitation, immediately reveals a further great region of which we are ignorant – 'Rationem vero harum Gravitatis proprietatum ex phaenomenis nondum potui deducere' (But I have not been able to discover the reason for this property of gravitation from the phenomena), and the same is true to-day.   The whole of the knowledge we have ever obtained through the medium of our senses still presents the same picture as Newton saw in his old age; a few beautiful pebbles and shells picked up on the shore of the great ocean of the Unknown.   But the very fact of the astonishing interconnection between everything that we are able to fish up from this ocean, and so come to understand more closely, assures us that the mind of mortal man has the right to take refuge in this ocean, being itself a part of it, and drawn to it when human fears must be overcome, and when humanity strives to understand its own existence.

It is worthy of remark here that Newton's complete humility in face of the great unknown, together with the

[1] This appears particularly from Newton's letters to Doctor Bentley, a theologian who applied to him in regard to these questions.

majesty of that which he brought to our knowledge, in no way produced a similar state of mind in the next generation; rather the very opposite was the case. We only need think of the 'Encyclopaedists' (Diderot, d'Alembert and others) and also of Voltaire. This obviously depends in part upon the fact that Newton presented to the mathematically gifted, and those with literary powers – both of which types are much more common than those with a gift for science – a great deal to do and to work out, that was already firmly founded in his own achievements. This led to a high regard being paid to those who performed these functions, whereas their sphere of activity was practically devoid of any contact with nature, which always keeps the true investigator humble by reason of the mass of the unknown hidden in it. It thus came about that the first main result of Newton's magnificent achievement was arrogance. Besides, in this period of 'enlightenment,' the untenability of a great deal of the moral sciences, especially theology, was rendered evident by the sudden advance in natural science due to Newton. Theology was supposed to supply the people with religion (that is, connection with the world of spirit); in this it failed for want of agreement with the unity of reality, and the result was a general uprooting of men's minds.[1] This process is not yet concluded even to-day; it results in materialism, which sees the main objects of the successful investigation of nature in mastering nature and in improving technology; and furthermore, allows these successes to serve, not the good of the people, but as a basis for greed of gain. Before matters got as far as this – the age of steam and then electricity – new investigators were still to come, who were again quite humble-minded, and discovered many more and quite different things. But it is noteworthy, and in conformity with

---

[1] We only need to remember that both the Popes and Luther had been opponents of well-founded natural knowledge since the time of Copernicus. It is also characteristic in this connection that Diderot and Voltaire were both educated by the Jesuits.

the uprooting, that the period of 'enlightenment,' at any rate
the beginning of it, coincides with an almost century-long
pause in great scientific progress.   This is the time from the
appearance of Newton's *Principia* (1686) to that of Watt on
the one hand, and Coulomb, Galvani, and Volta on the other,
who (all round about 1780 and 1790) created the foundations
of the new development we have mentioned.   For this however, remarkably enough, the development of new means or
new tools of knowledge was in no way necessary, but merely
the re-adoption of modest and patient devotion to nature.
This is also true of the discoveries of Scheele, Priestley, and
Cavendish (in the years 1770 and 1780), who founded the
new chemistry.

An important and outstanding part of Newton's investigations was his discoveries concerning the colours of light.
Here he appears entirely as an experimenter, as master of the
art of observation, who puts carefully considered questions
to nature by means of lenses and prisms.   The only previous
example, although in a quite different direction, of experimental activity so wide in extent, and so comprehensive and
illuminating in its effects, was the work of Guericke.   A great
deal had been found out about light from the time of Euclid
to that of Kepler, and so on to Huygens; but the nature of
colour, the relation of coloured light to white light, the development of colours in the rainbow, indeed everything in
which we have to do not with light alone, but with colour,
had still remained incomprehensible.   The tendency was
to regard white light as the original thing given, as by the
sun, and as simple in its nature, and coloured light as compounded of white light and something which colours it,
something coming from a coloured body and acting upon the
white light.   Newton started from the colours which result
from white light without the assistance of coloured bodies, as
in the waterdrops of the rainbow, or in the colourless glass
of a prism used to refract light.   By adding experiment to

experiment in a well considered manner, and conclusion to
conclusion – as we may read in the first book of his *Opticks* –
he arrives at the proof of the compound nature of white light,
and the existence of certain mono-chromatic lights which
cannot be further decomposed, as they appear in the rainbow
or in the prism, passing from red through yellow, green, and
blue, to violet. These when mixed give white, and are contained, already mixed together, in the original white light.
When refraction occurs, as in the raindrop or the glass, the
various colours separate, forming the spectrum, and they
separate on account of their own different degrees of refrangibility. 'Light rays of different colour are also of different
refrangibility' is one of the principal results of experiment
in Newton's *Opticks*.

Every such sentence is always followed by the 'proof by
experiment,' mostly by whole series of experiments varied in
all sorts of ways, such as to-day also form the introduction to
scientific optics. An important matter is the proof that ordinary coloured bodies are in no way capable of changing the
colour of light rays, but that they merely act selectively, so
that red paper for example, simply has the property of reflecting more red light than light of other colours which falls
upon it simultaneously, but not of making, say, red light out
of green. If the light of the sun, therefore, were mono-
chromatic, say red, as found in the spectrum, no other colour
would be visible; all bodies would then only appear more or
less bright red, down to black. Only one hundred and fifty
years later did Stokes find, in 'fluorescent' and 'phosphorescent' bodies, cases outside the range of Newton's law;
these bodies are actually able to change the colour of light.[1]
Newton then also showed that all sensations of colour whatsoever experienced by the eye, may be produced simply by mixtures of the pure colours from red to violet occurring in the

[1] Stokes lived between 1819 and 1903, and was professor of physics in
Cambridge.

spectrum; thus purple, for example, is a mixture of red and violet. He also gives a rule, the colour circle, by means of which the results of such mixtures may be predicted approximately. It is noteworthy that he already recognises how imperfect the eye is (without the assistance of a prism) in judging mixtures of colours; for example, the eye perceives as 'white' mixtures in which by no means all colours are present, as they are in ordinary white light; fewer colours are sufficient when mixed in suitable proportion, to produce the sensation of white in a way indistinguishable from the ordinary sensation.

These results of Newton's, stated so clearly with so much detail, and with admirable caution, long suffered a peculiar fate: they met with very sharp opposition. Best known from later years is perhaps Goethe's appearance on the field with his colour theory; but even immediately after the appearance of Newton's publication many objections were made. As is often the case in the history of science, when fundamental work, which brings something quite new, appears, simple misunderstanding is responsible. Such misunderstanding frequently depends upon loose terms, which everyone imagines to understand when they are used, but nevertheless does not understand. A bad term of this description is the word *colour*; it has at least three different meanings; 'red' for example may mean: (1) red light, that is the ray of light which when falling on the eye produces the sensation of red; (2) red may mean this sensation itself, the sensation of colour in the observer; (3) red may also mean a red colouring matter, that is a substance which reflects or transmits red light better than other kinds of light, and hence makes red light out of white light. It is obvious that the scientist who wishes to explain the nature of light, will, like Newton, mean the first, which by the way, Newton himself states with complete clarity.[1] It is equally true that the artist like Goethe, who in

[1] *Opticks*, book 1, part 2, prop. 2, definition.

the first place is interested in the living man, will generally think of colour in the second sense.[1]

The painter will generally take the third meaning as a basis. When he mixes yellow paint (for example gamboge) with blue (for example prussian blue) he gets green. But yellow and blue light when mixed give a more or less pure white. The white light thus obtained is identical for the colour-sensitive eye, and therefore in the second meaning of the word 'colour' above, but by no means in the first meaning of it, with the white of daylight, the latter being a mixture of all colours (in the first sense of the word). These examples serve to show – we have not space here for further discussion – into what confusion we may fall with the word 'colour' if we do not start, as did Newton in his *Opticks*, from a clearly stated and sharply worded definition.

The questions of colour have all of them been set upon a firm foundation by Newton's experiments. This was done by observations in experiments which, in part, were not entirely new, but which were used by Newton to draw new conclusions very general in range, and these led him to important developments of knowledge.[2] From what we have already said, the reader will have formed some conception

[1] We note when reading Goethe's theory of colour, that he does not succeed in grasping what Newton meant by colour. Goethe cannot get away from colour *impression*, and hence does not realise that in order to understand the matters in question as far as possible, it is first necessary to investigate that which *produces* the impression, that is the light itself, that which passes through space in the form of a ray and possesses properties which are quite independent of the presence of an observer's eye. This is what Newton was investigating. Scientific research must always proceed from the simpler to the more complicated, and not conversely. It is clear from much that Goethe says in his theory of colour, that we are wrong in regarding him as a scientific investigator; he was a great friend and spectator of nature, and that is much, when after the manner of Goethe.

[2] Markus Marci in Prague described in about 1648 experiments with prisms and lenses which in part were similar to those of Newton, without however arriving at clearly defined and regular results, upon which further construction could take place; also, he appears to be ignorant of the important law of refraction due to Snell (see E. Mach, *The Principles of Physical Optics*, trans. by J. S. Anderson and A. F. A. Young, 1926).

of the confusion upon which Newton threw new light. The fact that his contemporaries did not realise the existence of such confusion, and did not therefore admit it (this, indeed, did not happen until much later) caused Newton much uneasiness, which went so far that at times he was inclined to abandon his scientific activities.[1] For the Royal Society, to which he first communicated his results, required in every case that a reply should be given to objections and opposing views communicated to it, of which in this case there was a rich harvest, and Newton usually found it very much more trouble to disentangle the trains of thought of his opponents, than to make fresh discoveries in nature. If we examine the literature in question to-day, we get a strong impression of its complete uselessness; no new knowledge resulted. We thus again see quite clearly that great advances only come from single personalities, and not from societies, no matter how excellent the persons may in general be of which they are composed. Such societies should there-fore see their province exclusively in protecting and for-warding the work of the single and all too rare individuals who show themselves to be bringers of progress in any direction.

It was not however only the question of colour which brought what Newton felt to be embarrassing complications, but also the question of the nature of light generally. New-ton had made careful investigations concerning the peculiar colours shown by 'thin plates,' such as soap bubbles, or the layers of air between two surfaces of glass pressed together, and here also he made fundamental discoveries. The coloured rings which are seen under an ordinary lens when it is laid upon a plane glass plate, have since been generally known as 'Newton's rings.' In summarising his

[1] He regrets in a letter written in the year 1672, in one of the connec-tions which follow later above, that he has sacrificed such an important element of happiness, as his peace, in order to chase after a shadow.

account of the phenomena, he concludes that along the path
of a light ray there must be states, or 'fits' as he calls them,
of periodic change of different kinds, which he regards as
'fits of easy reflection and easy transmission,' and he is able
to measure the very small distances apart of these states,
and finds it greatest for red light and smallest for violet
light.   This is the fundamental fact which he deduces from
observation in a great variety of experiments.   He remarks:
'What kind of action or disposition this is; whether it
consists in a circulating or a vibrating motion of the Ray,
or of the Medium, or something else, I do not here
enquire.'[1]

Further advance was actually only assured by quite new
experiments made a hundred and thirty years later by
Fresnel.   Newton carefully avoided dealing with questions
which were not directly connected with a comprehensive
and clear description of the phenomena he had observed.
In particular, he did not propound any theory of light.
Quite in opposition to what Newton himself plainly says in
the second edition of the *Opticks* in 1717, we quite commonly meet even to-day, and the assertion that he did propound a theory.   He did not even set up any definite
hypotheses concerning the nature of light.   He says at the
beginning of the first book of his *Opticks*: 'My Design in
this Book is not to explain the Properties of Light by
Hypotheses, but to propose and prove them by Reason and
Experiments.'   In the second book, in which the colours of
thin plates are especially dealt with, he describes a series of
twenty-four experiments, and follows these by notes and
analogies, and then gives thirteen further related observations.   In the third book, which he expressly states in the
preface to be unfinished, he brings still further observations,
which later became of importance, on the diffraction of light,
and closes with thirty-one questions concerning a large

[1] *Opticks*, book 2, part 3, prop. 12.

number of matters which he regards as worthy of consideration.

In spite of these facts, we nevertheless find all too often even to-day, and even in text books of optics, the description as an historical event of importance, of a controversy between the 'emission or emanation theory or hypothesis of Newton' and the 'undulatory or vibration theory or hypothesis of Huygens'; this must be referred to the feeble mental grasp of Newton's and Huygens' contemporaries, and also to the carelessness of their successors, who for the most part became acquainted with these great men at secondhand, along with polemical statements introduced by others. For Huygens likewise attempted simply to bring facts together, as many as possible, which appeared likely to lead to an insight into the nature of light, and this is also quite generally the manner in which our knowledge of nature gradually comes into being. It has never been furthered by mere opposition of different suppositions, but always by discovery of new facts by means of suitable experiments and observations. It is true, however, that narrow minds are not able to keep the living facts in front of them, and are also unable to regard science as something which is continually progressing; hence for their own part they feel the need for schematising what is already known, with the result that it too easily becomes fixed in a rigid and onesided manner. They should not ascribe this mode of thought to the great scientists, who so obviously did not possess it, and they should not obscure the essential in the progress of knowledge by emphasising controversial questions; these have always been fruitless, since they amount to an 'either – or' (*tertium non datur*), which has nothing to do with nature.

The case of light, as the result of new and in part fairly recent discoveries, has proved to be an excellent illustration of this fact. Light is certainly a wave motion in the ether, as Huygens already imagined – this however also includes the

changing conditions in the ray which Newton dimly visualised – but it has astonishing and unexpected peculiarities
(namely it is transverse and electromagnetic in character,
and not as Huygens first tried to regard it, longitudinal and
elastic). Nevertheless, this in no way contradicts the old
idea, which Newton also expresses in some places, of a
something projected with the velocity of light, inasmuch as
every wave is moving energy, and energy always possesses
both mass (inertia and also weight), quite as much as any
material body. How much more wonderful is nature than
we ever suspect before we come to know! All great men
of science have been sensible of this; we feel it in reading
their works. But most ordinary writers do not do them
justice.

Newton's life history is quickly told.[1] He was very
comfortable in England. He grew up under the care of his
mother and grandmother, since his father died before his
birth. He proved unfitted to look after the family estate;
but they were also not able to do much with him at school.
It was great good fortune that in spite of this, and apparently
ill-prepared, he was able to enter the University of Cambridge at eighteen. His original tastes for mechanical
occupations,[2] for drawing, and for reading, then quickly
developed, by way of his mathematical studies, in the
direction where his lifework was to lie. He worked on his
own account on Descartes' geometry and Kepler's works;
Euclid soon appeared to him as self-evident. In this early
period also fall his own first experiments with the prism on
the nature of colours, his first ideas concerning gravitation,
and the beginnings of the calculus of fluxions, all of which
can be traced back to about the year 1666; they were then

---

[1] A biography of him by David Brewster appeared in 1831; see also the
works by S. Brodetsky and V. E. A. Pullin, both published in 1927 in
London.
[2] We are told of all kinds of mechanisms which he built, but which
were later all lost.

gradually developed further.   His communications on these matters to his university teachers[1] resulted in his being made the successor of one of these, and thus professor of mathematics at Cambridge in 1669 at the age of twenty-seven. He held this position actively until 1695, and all the achievements which we have described above fall in this period, which lasted to his fifty-second year.

Right at the beginning he also gave lectures concerning his optical discoveries. As a result of these, Newton also came upon the idea at that time, that the production of colours by refraction stood in the way of improving telescopes in which lenses were used, and this led him, in 1668, to make the first reflecting telescope.   This excited general interest, so that after it had been tested by the Royal Society and a description sent to Huygens, it was also shown to the King, Charles II.[2]

The fame of the telescope soon caused him to be made a member of the Royal Society in 1672, and he remained closely connected with it during the whole of his life.   Newton's consideration of the motions of the planets, which already began in 1666, led him, on the basis of Kepler's laws, to assume a power proceeding from the sun and inversely proportional to the square of the distance.   An identical force must also proceed from the earth and keep the moon in its path, and this force might – if everything were as simple as possible and nothing else came into play – even be the same as the ordinary force of gravity of the earth, reduced in proportion to the square of the distance.   One of the most important ways of testing this idea of universal

[1] The beginnings of the calculus of fluxions were already communicated to wider circles in 1669, and later also in letters by Newton himself. (Thus in 1679 to Leibniz – see the preface to the *Opticks*); the complete publication of the calculus of fluxions only occurred in 1736, after Newton's death.

[2] This telescope, or a second apparently likewise built by Newton himself, is still carefully preserved in the library of the Royal Society; it bears the date 1671.

gravitation was to see whether it agreed numerically with the radius of the earth and of the moon's orbit, whereby however, it remained to be proved that the force is to be reckoned from the centre of the earth, that is to say, that gravitation acts through the whole mass of the earth, even from the most distant part of it, quite undiminished. According to what was then known concerning the earth's radius, the agreement was only approximate, and this led Newton to lay the idea aside and devote himself entirely to his optical and mathematical studies.

Not until the year 1679 did he again take up the mechanics of the heavens, but he then continued to work at it with the greatest energy. Apparently the occasion for this was furnished by the expression by members of the Royal Society of various ideas, which obviously were linked with Kepler's laws and also with Huygens' laws of centrifugal force (published in 1673), and had already led to the discussion of a law of inverse squares.[1] Newton then, after a new measurement of the earth's meridian in 1682 had furnished a secure basis for the above mentioned calculation with respect to the moon, put forward in place of suppositions and disconnected statements, his well-constructed *Principia Mathematica Philosophiae Naturalis*, the rich contents of which we have already admired. The work was finished at the end of 1684; it appeared in print in 1686.

In the year 1685 an event of importance in Newton's life occurred. King James II desired to bring Catholicism into power in England, and for this purpose wished to have a monk of no particular capabilities accepted as a graduate

---

[1] In particular Hooke, and Halley (the first eight years, the latter thirteen years younger than Newton), and also Wren, may be mentioned as likewise busily engaged in considering the origin of the mechanics of the heavens; there can be no doubt that Hooke was rather a hindrance to Newton, whereas Halley was a help, inasmuch as he continually urged Newton to publish the *Principia*, and was also a close friend in other ways. Halley's assistance, particularly as regards the comets, has already been mentioned above.

of Cambridge University. The University refused. But when the King's command was repeated with threats, it was already near to giving way, but nevertheless formed a committee from among those of its members who stood firm, to defend its rights. Newton was also elected to this committee, and it was successful; the King retracted his command. A further result of this was that Newton was elected, in 1688, Member of Parliament for Cambridge University, but he only sat for two years.

Newton's Cambridge professorship did not make great demands upon him, and gave him the freedom necessary for his extensive researches; but it also only afforded him a very modest livelihood. It appears often to have happened in England at that time that men distinguished at the University were given another and higher field of activity in their later days. So Newton was offered in 1696 the position of Warden of the Mint, being regarded as particularly suited for the post, not only as a person of established reputation and a mathematician, but especially as a metallurgist, by reason of his numerous experiments with alloys for making telescope mirrors, and also in many other and varied chemical matters.

In this position, which was important at the time on account of the reformation in the coinage which was taking place, he was able to perform important services. The change in the coinage was completed in two years, whereupon Newton was made Master of the Mint, which position he retained for the rest of his life. In the meantime, he was able to supply a deputy for his Cambridge professorship, until he gave it up in the year 1703. The leisure afforded him by his position as Master of the Mint, was used for supervising the republication of his work, and for historical study, which however he did not intend for publication.[1]

---

[1] Such studies were in part published during his life against his will; their full publication only occurred after his death.

He also always remained an active member of the Royal Society, becoming its President in 1703, and remaining in this office until his death.

Newton was the first great and comprehensive investigator in history, to whom age brought no modifications of the convictions which he had expressed during his life. Although he had still many battles to fight, since doubts arose concerning his priority in the invention of the method of fluxions – doubts which he would not allow to be held as being contrary to the truth – he nevertheless had everyone on his side in his own country, and his position as Master of the Mint ensured him a comfortable old age. His household, since he himself did not found a family, was conducted by his married niece; he kept a carriage and three male and three female servants. He was hospitable and amiable, and delighted in great generosity.

Newton was not above the middle height, and somewhat stout in his old age. He had long and wavy hair, which was vigorous and silver white in his old age, so that he did not wear a wig. In his eightieth year he became infirm. He died at the age of eighty-five, and was buried in Westminster Abbey with all the honour due to one of the greatest of the land. A fine memorial with a suitable epitaph is dedicated to his memory.

GOTTFRIED WILHELM LEIBNIZ (1646–1716)
DIONYSIUS PAPIN (1647–1712)

LEIBNIZ's personality and life cannot here be described as completely as those of Newton and Guericke, although he has much similarity with these, since his achievements as a scientist are in the main indirect, concerned as they are with the great tool of scientific investigation, mathematics.

He developed the calculus of fluxions or infinitesimals. This method of calculation, which was specially designed for the prosecution of scientific investigation, and has become fundamentally important for it, deals with infinitely small quantities; Leibniz gave it the construction and form which has remained fully adequate until to-day. In this mathematical achievement lies his relationship to Newton; to Guericke he is related by his extensive services to his country,[1] and the latter were to him, as for Guericke, decisive as regards his external life, his undertakings, and his manifold travels.

Leibniz was born in Leipzig; his father, whom he lost at the age of six, was professor of moral philosophy at the University there, and at the same time a notary. In his earliest years he already began to read in the library of his father, far in advance of his school Latin, the old historians and classics with great enjoyment, and at fifteen he entered the university of his birthplace. Here, and then at times also in Jena, he first turned to mathematics; Descartes and Euclid were his studies, but he soon decided upon practical law as a career. At nineteen years he obtained his doctorate in law, at Altdorf, near Nuremburg, since his youth caused difficulties for him in Leipzig. As in his schooldays, so at the University, he made great progress, mainly by thinking for himself and using books chosen by himself, in everything at that time called science; he always arrived very quickly at the point at which he was able to go beyond the limits of existing knowledge on his own account, and he valued nothing more highly than the power and freedom to do so, as opposed to the advice frequently given that one should not attempt to do something new in matters which one has not yet studied fully.

---

[1] The latter are fully described in *G. W. Leibniz*, by Pfleiderer (Leipzig, 1870). There is an English biography of Leibniz by J. M. Mackay, published in 1845.

GOTTFRIED WILHELM LEIBNIZ

DIONYSIUS PAPIN

JAMES BRADLEY

By an essay on the reform of the theory of law, he attracted the attention of the Kurfürst of Mainz, who called him in 1668 to do legal work, whereupon his extraordinary versatility and power of work immediately led to his rising to continually higher posts in the service of the State. At this time, at the age of twenty-four, he was described in a letter of recommendation by one of his patrons as 'learned to a degree beyond anything that can be said or believed . . . industrious and fiery; independent in religion, but otherwise a member of the Lutheran church.'[1] At that time, it is true, State service in Germany was the service of small States, but Leibniz made higher plans on his own account. He wished to keep Louis XIV, who at that time was already getting ready to take Strasbourg, well away from Germany, and for this purpose secretly worked out a memorial for this mighty prince which was to recommend to him the idea of conquering Egypt. The memorial was to be handed over in Paris, and Leibniz, provided with good introductions, carried out the project as a senator of Mainz, in the year 1672; the desired success was not attained.

On this occasion, Leibniz found himself introduced into the highest circles in Paris, to which his published writings of various kinds no doubt also contributed. Here he also made the acquaintance of Huygens, which had long been his desire; otherwise he buried himself, according to his own statement, in the libraries. He thus came to know of a calculating machine invented by Pascal, which however only added and subtracted, and this he immediately perfected (no doubt utilising in part ideas which he had previously conceived) so as to make it execute multiplication, division and the extraction of cube roots. The machine was constructed in Paris with the assistance of the minister Colbert, and was recognised by the Academy; its construction served

---

[1] See, also as regards other matters, Guhrauer's biography of Leibniz, Breslau, 1842.

Is

as a model for the further development of calculating machines up to the present day.

In the next year, 1673, Leibniz went for the first time to London. There the secretary of the Royal Society (Oldenburg, born in Bremen), quickly introduced him into the circle of the Society, to which he also showed his calculating machine brought from Paris. Boyle, one of the oldest members, gave Leibniz special opportunities to exchange ideas with the mathematicians of the Society, whereby his extraordinary gifts, in spite of his own very modest statements (also in writing to Oldenburg) concerning his then deficient mathematical schooling, appeared to have been correctly estimated; for he was immediately made a member of the Society. A year before, Newton had been made a member.

After two months stay in London, Leibniz went back to Paris, where he remained three years. There he carried out extensive mathematical studies, having in this the help of Huygens, whose *Horologium Oscillatorium* formed the subject of his special study. He also corresponded with the London Royal Society, exchanging ideas on mathematical subjects. Leibniz thus moved in the circles which represented, in the person of the most eminent members, all ideas of that time concerning the necessity for calculation with infinitely small quantities which had arisen in the progress of science since Galileo's time; indeed, ideas concerning actual methods for this purpose, for Newton had already been for ten years in possession of one in his method of fluxions. It is certainly characteristic of Leibniz's quite extraordinary capacity for grasping the essential, and for immediately advancing to new results in all subjects that he took up, however little he may previously have concerned himself with them, when we see from the correspondence of that time that at the end of his stay in Paris, in the year 1676, he was already in possession of the new calculus in a fully developed state, calling it the 'differential calculus,' without anything having been published about it from any side up to that time. Leibniz also contented himself, like Newton before him and as was common among learned men of the time, with communications to smaller circles, to London and of course in Paris; his first printed publication of the differential calculus, 'a new method for maxima, minima and tangents, which avoids fractional and irrational quantities, and a peculiar method of calculation relating to these,' appeared in 1684.

Here we have for the first time the method, which quickly became general, of describing infinitely small quantities (differentials); the rules of differentiation are developed, which to-day belong to the fundamentals of higher mathematics. This terminology was the secret of the easy applicability of the new method of calculation. Leibniz says himself on this point in a letter of 1678, that terminological expressions in mathematics are most helpful 'when they express the inmost nature of the matter shortly, and as it were give a picture of it.' 'In this way the labour of thought is reduced in a wonderful manner.' In actual fact, mathematics, as far as it serves natural science, has the task of keeping the labour of thought directed entirely towards discovering the inmost nature of things, and then of forming such an image of the discovery, that it is preserved faithfully and unfalsified by means of the rules of calculation, and can assume almost any desired variety of forms, which are necessary for its application in complicated cases.[1]

Leibniz was soon interrupted in the prosecution of his mathematical labours by again entering into active political work. The Kurfürst of Mainz, in whose service Leibniz had hitherto been, died in the year 1673. Three years later,

---

[1] It happens in the case of more recent applications of mathematics that the labour of thought has been directed not so much to the discovery of the actual behaviour of nature, as rather to arbitrary interference with the course of calculation, whereby the picturing of the inmost nature of the thing can only be falsified. This is obviously a misuse of the advantage of reduction of the labour of thought, since what has been saved is applied at the wrong point.

Leibniz accepted an invitation to the court of Duke Johann Friedrich, of Hanover, with whom he had already been in correspondence for some time, and this ended his stay in Paris. The return journey to Germany was made via England and Holland. When in London, where he only stayed a week, he inspected manuscripts of Newton, which contained matter relating to the calculus of fluxions. This was also before his first publication on this subject, and hence it cannot be doubted that Leibniz did not work independently of Newton, who was the first to conceive the idea of calculation with infinitely small quantities. But it is likewise clear that these ideas had gradually become known to many both from Newton and from Leibniz's remarks, and that nevertheless no one other than these two was able to give them permanent form; and further, that Leibniz alone gave them the final form which was so favourable for their application.

At the court of Hanover, Leibniz was first employed as Librarian, and as writer of the history of the royal house; but he was soon in request for higher matters of law and state. Along with these activities, and many other interests, he published during this period a further series of very important mathematical essays, relating to the further development and application of the differential calculus.

The last period of Leibniz's life brought him few satisfactions. It is true that he was able to witness the beginning of the rapid application and development of the differential calculus; but during his lifetime his services in this respect were not recognised. It appeared for a time, however, as if he would receive the major portion of the credit, since in his publication he did not refer to the assistance which he had received, apart from Huygens, particularly in the circle of the Royal Society, whereas Newton openly recognised[1] that

[1] *Principia, lib.* 2, *sectio* 2, *lemma* 2, *scolium* (1686, two years after Leibniz's first publication concerning the differential calculus).

Leibniz had, in a letter to him in 1676, already communicated to him something concerning the nature and range of his new method of calculation; but the position was soon reversed, since the Royal Society claimed all credit for the first originator, Newton.   Contemporary judgments of this kind must necessarily rest upon the opinion of outside persons, that is of persons who not only did not make the discovery or invention, but were actually remote from it, and therefore not capable of making it, or at any rate did not follow the development of the ideas in question, and hence were not capable of forming a judgment.   The consequence of this is that contemporary judgments concerning the credit for great advances must always be worthless, the more so because even the knowledge of what such achievements depend upon, and what makes their excellence and enhances their value, is in any case only accessible, by inward experience, to the few who themselves belong to the ranks of the great investigators.   Posterity can more easily form a correct judgment; it already possesses and controls the ideas, about the origin of which a judgment is to be formed, and it also has at its disposal more complete means of following back all threads leading to the origin.   But contemporaries can form a true judgment of the living investigator from his manner and style of work, and they can recognise unusual powers by unusual achievements, whereby, however, a satisfactory judgment is always only possible between men of like nature (race).   For this very reason every man of science ought to, and should, find deserved recognition first of all among his own people.   This was not the case with Leibniz. He was certainly, when young, highly regarded by not a few German princes, with some of whom he was even on terms of personal friendship, but when these princely patrons of his, whom he served with love and devotion, were dead, he found little consideration, and indeed at the last neglect from their successors (George of Hanover, who became King of England in 1714); cared for only by his secretary, he died at the age of 70, a lonely man. Not until fifty years after his death was his grave in the Court Church at Hanover made recognisable by a suitable inscription.

Leibniz is described as of middle height, stooping somewhat, with broad shoulders but spare of build. His hair was very dark.

Leibniz's many travels, his relationship to the princes, his great versatility, and his extensive correspondence, gave him great influence in the scientific life of his time. He became in 1700 founder of the Berlin Academy. His correspondence, particularly with Papin, shows him concerned in many directions with inventions which only later became practicable, as for example the proposal to make a barometer without mercury, consisting of a flexible closed vessel with a vacuum inside and held by a spring in equilibrium with the atmospheric pressure; the aneroid barometer of the present day. He was also concerned with the steam engine, as the following facts go to show.

Dionysius Papin, the first to conceive a steam engine with a cylinder and piston, and to construct one, was born in Blois (south of Paris), and studied medicine. At the age of twenty-four he went to Paris, where he made the acquaintance of Huygens, to whom he attached himself; and this brought him also in contact with Leibniz, with whom he continued to exchange ideas by a correspondence which at times was very active.[1] Huygens at that time was busy with air-pump experiments (he also introduced in the course of them the flat plate with the convenient bell receiver set upon it, which has always since been retained), whereby Papin proved a very gifted and skilful experimenter. The work of Guericke which had just appeared concerning the

[1] See Leibniz's and Huygens' correspondence with Papin (edited by E. Gerland, Berlin, 1881).

Magdeburg experiments, put the great powers of the atmospheric pressure into the minds of all investigators, and Huygens at that time designed a machine which was to pump the water out of the Seine for Louis XIV's fountains by means of atmospheric pressure. Papin carried out the machine and introduced it in the year 1674 to the minister Colbert. It consisted of a cylinder with a piston; this was first pushed upwards by gunpowder ignited on the floor of the cylinder, the excess of the gaseous products of combustion escaping by valves at the upper end of the cylinder; then, when the gases had cooled off, the air pressure drove the cylinder down, whereby it actuated the water pump through a rope running over a pulley. For every stroke, fresh power had to be introduced into·a small hold underneath the cylinder, and set on fire.[1]

Boyle in England had his attention drawn to Papin, no doubt by Leibniz, and invited him thither in the year 1675; he was to help him in his air-pump experiments and otherwise. Papin fulfilled this task in such a way, that in five years he was made a member of the Royal Society on Boyle's proposal. On this occasion he presented a paper on the 'Digester' which he had just invented, and which is still known under his name to-day, being used for cooking under increased pressure. In this Papin also introduced the safety-valve, which enables high pressures to be limited, and is to be found on every steam boiler. He also observed the dependence of the boiling point upon the pressure, although at that time the temperatures were estimated only by the time of evaporation of a certain quantity of water from a hollow in the lid of the pot. Soon afterwards, Papin accepted an invitation to Venice to found a scientific academy, but only remained there two years, returning to the Royal Society

[1] We see that this motor was the forerunner of the later gas-engines which were introduced in about 1873, which also operated by air pressure, and made use of a mixture of illuminating gas and air instead of gunpowder.

in London again, which now gave him the title of 'temporary curator of experiments.' His duties were to attend to the experimental demonstrations at meetings of the Society.

In the year 1685 Papin was to learn that he had lost his native country, since all protestants were banished from France (Revocation of the Edict of Nantes). Many went to Germany, and were welcomed there; thus the Landgrave Carl of Hessen-Cassel invited a relative of Papin, and finally Papin himself, to the University of Marburg, and Hessen thus became a second home to him from that time forth for almost twenty years. He also founded a family there. He lectured on mathematics and parts of physics; but his main efforts were directed towards realising several inventions with which he was continually busy, above all a variation of Huygens' powder motor, whereby hot steam was to be used instead of the powder gases, which offered, as he correctly realised, the advantage that a much completer vacuum would be produced after cooling the contents of the cylinder, and hence the air pressure would be made better use of; also no residue would be left behind. This was the fundamental idea of the present-day low pressure steam engine. The Landgrave also, when first he received Papin, showed great enthusiasm for such a machine, taking as he did a lively interest in all inventions of Papin, no doubt particularly because they were applicable to fountains, mills, and mines. Unfortunately, Hessen, like the rest of Germany, was kept in a state of perpetual warfare by Louis XIV, and this continually diverted both the Landgrave's interest and also the necessary financial means from Papin's undertakings, with the result that they were often greatly hindered; this was a repeated disappointment to Papin, the more so as he was without means of his own of which he could make use. Orders were also sent from time to time which meant breaking off the work, or giving it an entirely altered direction. In these circumstances Papin published his plan for

the steam engine in the year 1690 before it had been carried
out on an effective scale.[1] He there also formulated the
idea of the high pressure machine, in which the pressure of
the steam in the cylinder drives the piston, and the production of a vacuum by cooling is unnecessary, and he already
suggests the use of the machine for other purposes than
pumping water, for example, for driving ships by means of
paddle wheels. The piston rod was to be toothed, and act
upon a toothed wheel on the axis to be turned. A boiler
was not provided, the cylinder itself was to be alternately
heated and cooled.

In the meantime Papin, in 1692, devoted himself to
building a submarine boat, and in 1698 he turned his attention to the centrifugal pump which he had invented. The
submarine boat was successful, after one or two failures, to
the satisfaction of the Landgrave.[2] The centrifugal pump,
a first application of centrifugal force as studied by Huygens,
worked well with air, when used to ventilate a mine shaft;
but for continuous pumping of water, the man-power used
at that time was not sufficient for the very rapid rotation
required.

In the year 1698 Papin's circumstances improved, for the
Landgrave invited him, to his great joy, to be near him at
the Castle, in order to construct a machine to pump water
from the river Fulda to the top of a tower of the castle on
the river bank. Papin departed from the plans he had
published, and made use for this purpose of a special iron
boiler, the steam from which pressed either directly, or
through a piston floating loosely, upon the water, while
alternately, when the steam was cut off, fresh water flowed
into the space occupied by the steam. The machine was
unfortunately destroyed in November of the same year by

---

[1] *Acta Eruditorum*, August 1690.
[2] According to the drawings which have been preserved, Papin's
submarine may be described as an improved diving bell, closed at the
bottom.

the ice floating on the river, and this led also to the Landgrave's losing interest for the affair.   Pumps in which steam
acted directly upon the water had meanwhile come into use
in England.   Even thirty-five years earlier (1663) a pump of
this kind had been constructed by Lord Somerset, Marquis
of Worcester, with the aid of the skilful mechanic Kaspar
Kalthoff in London, and excited great interest, so that
Lord Somerset composed a special prayer for himself in
order to preserve himself from excessive pride over his
success.[1]

After being forgotten for decades, no doubt for want of
sufficiently skilful construction, this invention was taken up
again by the Englishman Savery, who in the year 1698 (the
same year as Papin) described a new and improved form of
the pump, and showed it in 1699 as a model to the Royal
Society, and later to the King of England.   Landgrave Carl
in Cassel heard of this and was thereby led to giving Papin,
in the year 1705, a fresh order to carry out a machine of this
kind.   It was finished in 1706 and pumped water to a height
of 70 feet; but the pipe line, which was cemented together
from a number of pieces, did not remain watertight.   The
Landgrave then had a pipe made of copper, but was continually hindered from attending to the matter, and no
experiments were allowed to be made without his presence.
Finally, in the year 1707, even the copper tube was taken
away from Papin, since it was destined for another purpose.
This and other similar misfortunes, finally led Papin to
carry out the plan he had always borne in mind, of again
seeking his fortune in England.   He first published in the
same year an account of his water pumping machine in a
special pamphlet which he sent to Leibniz, whereupon the
latter communicated to him the idea of automatic operation
of the valves, thus enabling the person who had to work the
stopcocks to be dispensed with.[2]

[1] See Poggendorff's *History of Physics*.   [2] *Correspondence*, p. 375.

Leibniz advised Papin strongly against moving to England; and the change turned out to be fatal to him. For the purpose of the journey he had constructed a small ship (capable of carrying four thousand pounds), which he provided with paddles, such as he had already seen in use in England when previously there; these were to be driven by hand. With this he intended to travel down the rivers Fulda and Weser with his family and property to Bremen, where the ship was to be sent over on another ship, since he intended to fit it in England with the cylinder and piston for use with steam.

Before his departure he showed the ship to the Landgrave, on the river Fulda. It was destroyed near Münden by the river sailors, who would not tolerate the presence of strange ships, and in England he found himself entirely without assistance. His one-time patrons Boyle and Hooke were dead; the Royal Society would not interest itself in the many plans which he laid before them. He died, obviously in great poverty, and finally lost to view, probably in the first half of the year 1712.

Though Papin's fate was sealed, the further development of the steam engine nevertheless started from his idea of replacing the powder gases in Huygens' piston and cylinder machine by steam, and Leibniz's proposal for automatic operation was also soon carried out, though it was re-invented. Hooke had given Papin's publication of 1690 to an iron merchant, and very skilful smith, Newcomen, who carried it out with great industry, and then joined forces in the year 1705 with Savery, who had already pumped water by using a simple steam boiler, without cylinder and piston, in the year 1699, as we have already mentioned. Thus in the year 1711 the first low pressure steam engine with boiler, cylinder, and piston, and soon also automatic operation of the valves, came into continuous service. Its reciprocatory motion served for operating the drainage pumps of a coal-mine, and more such engines were soon built for the English coal-mines. More trustworthy and efficient operation, and more general applicability through rotary motion, was first attained seventy years later by Watt, who added the new idea of a separate condenser.

JAMES BRADLEY
1692–1762

THE introduction of the Kepler telescope for astronomical measurement, of which Roemer was the pioneer, and a corresponding improvement in the accompanying divided circles, resulted in increasing refinement in the determination of the positions of the fixed stars. Particularly in view of the annual motion of the earth, astronomers had sought with growing hope of success since Copernicus' time for an apparent displacement, annually repeated, of nearer fixed stars as compared with more distant ones. This displacement is called the 'parallax' of the star, and its amount gives us immediately the distance of a star in terms of the diameter of the earth's orbit. Signs were also found of such displacement in the position of certain fixed stars, and Halley in 1718, was the first to determine the proper motion of Aldebaran, Arcturus and Sirius; an important discovery in itself.

But these motions were not periodic. Bradley was the first to use sufficiently refined methods and sufficient patience to discover annual displacements, the existence of which had also been asserted (for example by Hooke); he was the first, however, to determine them beyond doubt, so far as the available means at that time permitted, and he also interpreted them correctly. In the year 1778, he had been successful in making sufficient measurements of a star. It

appeared that the displacements had the expected yearly
period, but quite unexpectedly did not correspond to a parallax.   For the motion was not opposed to that of the earth;
on the contrary, the star appeared displaced every time in the
same direction as the earth's motion.

The following explanation was soon found by Bradley
for this unexpected displacement, which was then called
'aberration.'   If, as Roemer had observed, the light coming
from a star takes time to be propagated in the telescope, and
the telescope is moving with the earth across the direction
of the light ray, it will be necessary to incline the telescope
a little in the direction of motion, in order to receive the ray
along the axis of the telescope, and the star will thus appear
displaced in the direction of motion to this extent.   This is
the same phenomenon as when we move along under rain
falling vertically; we become wetter in front than behind,
as if the rain were coming from in front of us.   If this explanation is correct, the angle through which the star appears
to be diverted, must be given by the relationship between the
earth's velocity and the velocity of light, and this is found to
be the case.

This agreement meant new insight of the greatest importance.   Roemer's measurement of the velocity of light was
brought into an entirely new kind of connection with Copernicus' and Kepler's knowledge of the earth's motion; and
this amounted to a confirmation of all these facts.   At the
same time, the discovery of aberration gave us a new phenomenon of light, which will be of importance for all time as
regards questions of the ether, and of absolute motion
through space.

The angle of aberration is very small, in accordance with
the very great velocity of light; it amounts only to twenty
seconds of arc.   The fixed-star parallaxes originally sought
for, are still smaller; even for the nearest stars they amount
to less than one second of arc. It was therefore not until

one hundred and nine years later, after telescopes and divided circles had been again greatly improved by Fraunhofer, that the parallax of a fixed star was determined for the first time.

Bradley was born in the English county of Gloucester on the river Severn, and was professor of astronomy in Oxford from 1721 to 1742. He owed the means for his discovery to the private observatories of wealthy amateurs in astronomy. He afterwards became Director of the Royal Observatory at Greenwich, and successor of Halley, where he carried out in particular his observations on the nutation of the earth's axis, which had already been calculated by Newton.

JOSEPH BLACK (1728–1799)
JAMES WATT (1736–1819)

BLACK was the founder of the measurement of quantities of heat, or calorimetry.

The state of heat, temperature, had been measured by thermometers since the time of Galileo, at least in so far as that each investigator was able to mark upon his home-made thermometers points corresponding to definite temperatures. The temperature of melting ice was soon afterwards brought into use as a fixed point, assumed unalterable, in constructing thermometers. A second fixed point was given, when Papin had discovered the boiling point of water to be independent of the pressure, the new fixed point being the boiling point at an agreed 'normal' barometric pressure. Thenceforward, properly defined thermometric scales became possible; they were divided between the ice point and the boiling point into 80 (Réaumur), 100 (Celsius), or 180 (Fahrenheit) degrees.

But a quite different question remained, that of stating the *quantity* of heat – the unknown something – which is needed in order to raise the temperature of a given body by a given number of degrees. It had never been doubted that double the quantity of the same material requires twice as much heat; but what about the heat required for equal quantities of different substances? This question appeared very difficult, and even insoluble. At that time only a few superficial ideas were in existence. Black it was who created the necessary new foundation for dealing with all questions of heat quantity, by an extensive series of observations founded upon a new point of view. When he wished to compare a body as regards the heat needed for warming it, or the heat released from it on cooling, with water, he allowed the heated body to give up its heat to water, and determined the two initial temperatures and the common final temperature. Starting from the idea that the one body must have given up the same amount of heat as that received by the other, if transference of heat to or from the surroundings has been avoided, he was able to compare the capacity for heat of different substances with that of water, and hence also with one another, without inconsistencies. This is the complete fundamental idea of the mixture calorimeter, by means of which, even to-day, capacities for heat (later called 'specific' heats), are measured. Another calorimetric method, the method of cooling, was also applied extensively by Black for the first time; and he was likewise responsible for the fundamental notion of the ice calorimeter, which was later refined in the highest degree by Bunsen, over a hundred years afterwards, and became one of the most exact methods of measuring heat.

In the latter case, the foundation was given by a particularly important general truth discovered by Black, and entirely opposed to opinion at the time. He found that the change of state – melting or boiling – of a given amount of a substance does not only require a certain temperature, but also a certain quantity of heat, which is always consumed, though it cannot be detected by the thermometer. He proved for example, that when ice becomes water, the thermometer does not rise above zero, although heat is continually added until all the ice is melted; for this reason, also, the ice point is an assured fixed point for the thermometer. Black also measured the 'heat of fusion' of ice, and the 'heat of evaporation' of water (he called them the 'latent' heats), and his results are as near to the figures known to us to-day as could possibly be expected from the first practice of a new idea and method of measurement. 'It is scarcely possible to gain a deeper insight by attention to experiences by no means striking and accessible to everyone, than Black does here. Besides a receptive eye for the processes of our everyday surroundings, we have the acute analysis of each single experiment, and skill in the successful application of simple means.'[1]

Black began investigations in calorimetry previously to 1760.[2] In 1762 he communicated the results to a learned society in Glasgow, and thenceforward described them regularly in his university lectures, whence the ideas spread. A printed publication followed after his death, from his manuscript lectures (1803).

Black's earliest work related to another matter, in which

---

[1] E. Mach, *Die Prinzipien der Wärmelehre*, Leipzig, 1896, p. 163.

[2] Proof exists that in the year 1760 Black communicated his results to his contemporaries by word of mouth, for example, to Watt as regards the heat of evaporation of water, and its considerable magnitude. From the year 1772 onwards Wilke of Mecklenburg – who also rendered services to the investigation of electricity – published in the Swedish Academy, experiments relating likewise to consumption of heat in change of state, and in change of temperature of various bodies. He was the first after Black who pointed out the essential features of these questions. Lavoisier and Laplace only came forward with calorimetric work in 1780. Very noteworthy in this connection – and otherwise also – is the enthusiastic admiration shown by Lavoisier for Black in his letters, as compared with his avoidance of public mention of Black's name. (See the report of the Editor of Black's Lectures, vol. 3, pp. 21, 31.) Wilke is also only given the date 1781 by Lavoisier.

JOSEPH BLACK

JAMES WATT

he also brought forward matters of fundamental importance, namely the question of the mutual relationship of 'mild' and 'caustic' alkalis, as for example, chalk and lime. It was supposed that chalk when ignited gave off 'fire-stuff' (which as 'phlogiston' played a great but somewhat mysterious part in the thought of that period), and that it thereby became the corrosive 'quicklime.'. Black showed by thorough experiments that it does not take up anything that is perceptible, but on the contrary, gives up something when ignited, namely a kind of air; the same, namely that also results from fermentation (to-day called carbonic acid gas). He also proved that this production of gas is associated with a loss of weight. He recognised this kind of air quite generally as an essential component of all mild alkalis (such as unburnt magnesia, potash, soda), in which it is fixed in a solid form, and for this reason he called it 'fixed air.'

This was the first case in which a gas was recognised with certainty as a weighable constituent of a solid body. Also, for the first time the study of gaseous bodies was shown to be of importance. All sorts of varieties of 'air' had appeared in chemical experiments,but they were merely looked upon as ordinary air with something mixed with it, and not as special substances. At the same time, the idea of phlogiston, or fire-stuff, in the sense of a constituent of chemical compounds, became for the first time rightly suspect. Scheele, Priestley and Cavendish, and as a result of their work, Lavoisier, followed the matter up even during Black's lifetime, while Black himself approached the question of fire in quite a different manner, by means of the calorimetric investigations of which we have spoken.

Joseph Black was born in the south of France, but his family came from Scotland; he was one of thirteen children. His education and first training were received in Ireland; from the age of eighteen he studied medicine and natural science at the University of Glasgow, and four years later in

Ks

Edinburgh.   In the year 1756, he became Professor of
Chemistry at Glasgow as the result of his work on the
alkalis, as successor to his own teacher; ten years later he
succeeded him again in Edinburgh, and died there at the age
of seventy-one.   In his early years he practised with great
success as a medical man.   The fact that he published little
during his life, but communicated all his knowledge by word
of mouth, particularly in his lectures, which were very much
liked and which he left behind him in full manuscript form,[1]
is in accord with everything we know concerning his very
high character and the never changing calm of his mind. He
was very tall and thin, with a pale complexion; his large
eyes, we are told, were clear, dark, and deep.

James Watt, famous for his development of the steam
engine, who produced the low pressure engine by inventing
the separate condenser and designing all the other details,
and also carried out this engine on a large scale, was born at
Greenock, in the west of Scotland.   He was a very lively but
rather sickly child, who soon began to read greatly, but also
had a strong taste for mechanical occupations   By reason
of the poverty of his family, he was apprenticed at the age of
eighteen to a mechanic in Glasgow, and then sent to London.
Three years later he became mechanic at the University in
Glasgow.   His workshop was much visited by the professors,
who took pleasure in his versatility, his skill and his simple
and open nature.   Black also made friends with the mechanic
eight years his junior, who was so allied to him in nature,
and an intimacy between the two remained, which was
particularly shown in the support which later was given by
Black to Watt when in difficulties.

In the year 1759, Watt learned for the first time of the
steam pumping engine, called the 'fire machine,' which was

[1] They were collected and published, with a preface by John Robison,
in 1803.   See also Black's *Life and Letters*, by Sir William Ramsay, 1918.

rare at that time, and later that the University possessed a model of such an engine, which however was at the time in London for repairs. The model was sent for at Watt's request, to enable him to repair it: but it did not arrive until 1763. In the meantime, Watt had already made a number of experiments with a Papin digester as boiler, and with the piston and cylinder of a syringe, and had also learned all he could from books, whereby Black also helped him in conversation. He saw that the pressure of the steam, if the boiler were sufficiently heated, would be amply sufficient to drive an engine, without the use of the air pressure (which is the case to-day with all high pressure engines, for example locomotives); but at that time it was not to be thought of that boilers capable of standing the necessary high pressure in safety could be built sufficiently large; this was no doubt also the reason why Papin had already preferred the low pressure machine, in which the steam was only used to form the vacuum. Watt then soon set the model going, and he also saw the grave defects which had prevented these machines being popular; namely, the excessive consumption of steam. At every stroke of the piston, the cylinder, which had just been cooled, had again to be heated to the temperature of the boiler by the inflowing steam, only to be cooled again, which fact was quite sufficient explanation of the excessive consumption of fuel. While walking one day on a Sunday in spring 1765, Watt conceived the idea of the separate condenser. He perceived that it would suffice to connect the cylinder, which could then remain always hot, with a space always kept cool, after it had been cut off from the boiler, in order to obtain the required vacuum, since the steam would of itself stream into the cold space and there condense to water. He suddenly saw clearly the fact, now a commonplace, that in every space filled by saturated steam, the final pressure will always correspond to the coldest part of the space.

We may imagine how great was the progress represented by this notion as compared with the complete uncertainty concerning questions of vapour pressure, which Leibniz and Papin had found, until they themselves came to perceive that the pressure phenomena exhibited by heated water are to be ascribed to the latter and to its vapour, and not to an 'expansive constituent,' which it was supposed must be present in the water. It was thirty years later before Dalton was able to advance the question of vapour pressure further than Watt had carried it.

The separate condenser which Watt immediately tried out on a small scale proved at once satisfactory in the expected manner. With it Watt's low pressure steam engine was in the main invented; the other accessories necessary for its good performance, such as air pump, feed pump, condenser water-pump, flywheel, centrifugal governor, double action of steam on both sides of the piston, occurred to Watt directly – but not without his having to experience the length of the stride from the invention to practical production on a large scale. We are not able to follow his progress in detail.[1]

Papin had been in a like position, and had gone under; Watt had over him the advantage that Papin's ideas had already been carried out in a large scale in the machines, however imperfect, of Newcomen and Savery; while the further great advantage he possessed was that of being not only the inventor, but also himself the mechanic who carried out the work. Nevertheless, that did not help him as much as he no doubt imagined it would. It was no trifle

[1] The reader may be referred to the full account of his life given in *Lives of the Engineers* (Boulton & Watt), by Samuel Smiles, London 1878. The worst part of the difficulties which continued to threaten Watt, lay in humanity's greed of gain, which made the more difficulties for him, the greater the success of the machine. 'The rascality of mankind is almost beyond belief' was one of his remarks to Black. A later work *Watt and the Steam Engine*, by Dickinson and Jenkins, Oxford 1927, reproduces many portraits and historical machine drawings.

for the means available at that time, indeed, for a while it appeared almost hopeless, to construct the cylinder and piston of sufficient size and so perfectly formed as to be permanently steam-tight and yet able to move freely; all this was necessary in order to allow the value of the condenser to be felt, and the difficulty became greater still when the attempt was made to work with steam on both sides of the cylinder. Nevertheless, Watt devoted himself entirely to the practical realisation of his engine. He left the university in order to open a larger workshop, but was not able to maintain it, since the engine swallowed up money, without being able to show success. Black, who had already helped him out of his own pocket, but was not able to do so further, then brought him into contact with a well-to-do and sensible man, Roebuck, who wished to have a machine made by Watt at his expense for pumping the water from his coalmines, and who himself possessed the necessary workshops. Watt at that time maintained himself by occasional work as a surveyor. Nine years passed in this way, but the engine was not successful. The coal-mines were flooded, and the confiding man became himself as poor and as deeply in debt as Watt.

It was obvious that Watt's ideas could only be carried out, if at all, by means of the most perfect technical appliances available at the time, such as were only found in a few very large workshops in England, but not in the poor Scotland of that period. The possessor of such a workshop, in Soho near Birmingham, Matthew Boulton by name, was a very enterprising person, at the same time full of interest in scientific matters, and a man of very fine character. He was ready to join forces with Watt; so the latter left his own country and went in 1774 to Birmingham, to which place also the most valuable parts of the experimental machine already built were brought. After two years, the first engine was at last completely successful; it was destined to drive the air-blower of a blast furnace. It was a day of re-joicing such as Watt had not known for long, when this success was obtained, and the joy of his old friend Black, who had always followed his work sympathetically, was also great. The news of the engine's powerful and reliable action, and its low consumption of fuel, soon spread, and the number of engines built according to Watt's design continually increased. Many coal-mines could only be operated at all after Watt's engines were available to pump them dry, and he himself was almost always obliged to supervise the in-stallation in order to make sure of success. Only after six years, 1782, did he find time to construct engines with a rotating flywheel. It is noteworthy that he was obliged to avoid the used of the crank already common to every lathe, for converting the reciprocatory motion into a rotary motion, and had to develop a special mechanism for this purpose (the 'sun-and-planet gear'). Someone had guessed Watt's intention, and obtained a patent for steam engines with cranks. With the introduction of the flywheel, the development of the engine on a large scale was completed, and it remained without the addition of any new idea of importance for over a century. For the high pressure engine, the expansion engine, and the tubular boiler, had already been thought of by Papin, Watts, and Boulton, and the much deeper theoretical foundation later given by Carnot and Robert Mayer, did not add anything essential to the construction of the machine. Watt had at first hesitated to suggest that even mills and other factories, which were already in existence driven by waterpower, might go over to steam and work on a much larger scale, but this soon occurred, since coal, thanks to the pumping engine, could be obtained easily and cheaply everywhere. The result was a revolution in all mechanical factories, in particular a very rapid advance in spinning and weaving, since it was soon possible to employ hundreds of horsepower with certainty; the age of steam began. Watt was able to experience the promising beginning of all this, and as he grew older he was able to rejoice in greater and greater success; even the headaches from which he had suffered greatly during the period of his highest activity, at last left him, and he enjoyed a comfortable old age. He had married twice – his first wife died early – and he lived to the age of eighty-three.

We must not forget that Watt also took part in pure scientific research. As a friend of Priestley, he was concerned in the discovery of the composition of water which was then in progress, and was completed by Cavendish; water had hitherto been regarded as an element.[1] Very notable is the complete clarity with which he deals with the concept of work, in which he makes a considerable advance upon Leonardo and Stevin; however, in the development of the steam engine, this must appear almost self-evident. He measures the work done in the cylinder of his engines quite rightly by the product of pressure and volume, which here takes the place of the product of force and distance. For this purpose he constructed a special instrument, which could be connected to the cylinder, his so-called indicator. Also, the measure at present used for work done in unit time, the horsepower, was introduced by him.

Watt received an impressive memorial in Westminster Abbey; the inscription praises him as benefactor of humanity. He certainly deserves the term as far as he was concerned. He, as the inscription says, 'increased the powers of man.' They are certainly increased ten times to-day by means of the steam engine. But why do we not find that man has become happier and on a higher spiritual level through Watt's present to them, which enables natural force when

---

[1] The reader may be referred for an account of Cavendish's work to the *Life*, by G. Wilson, 1848; also to the *Collected Papers*, Cambridge, 1921.

properly used to do what the muscles of man himself and of his domestic animals could never have done? We may equally well ask where are the enterprising people who set up steam engines in order to make people happier, and not in order to gain the financial profits upon which they have calculated. Watt made other use of the gifts of his mind, than these minds have made of his gifts.

WILHELM SCHEELE (1742–1786)
JOSEPH PRIESTLEY (1733–1804)
HENRY CAVENDISH (1731–1810)

THESE three great discoverers, all enthusiastic and exact experimenters, founded, together with Black, the science now called chemistry. This happened by the discovery of a large number of facts hitherto unguessed at, which dealt with the commonest substances, such as water and air, and by the discovery of new substances, which although they had always been present around us everywhere, and had always played a very important part, had not yet been separated and examined in a pure state.

Using their present-day names, these were hydrogen, oxygen, chlorine, which, with their peculiar new properties, suddenly appeared like a new world full of marvels; at the same time much light was thrown on many matters, such as combustion and breathing, which had hitherto remained completely incomprehensible. Furthermore, not a few of the newly discovered bodies were gaseous, and bodies of this kind had hitherto been known simply as 'air,' whereas these investigators showed that there are just as many fundamentally different kinds of air (for which the word 'gas' now came into use), as there are liquid and solid

bodies, and that these gases take just as essential and weighable a part in chemical transformations as do other substances.
Black had already given us the first example of this, in
carbon dioxide.

Thus the gaseous state of matter first became recognisable
as of full and general importance. Gases thenceforth
became continually the object of ever new investigation,
since these three workers had also developed the necessary
methods for dealing with them. The chief of these was
their enclosure in glass vessels over mercury, which neither
dissolves them as does water, nor contaminates them by its
vapour.

The three scientists cannot be separated from one another.
They possess – as did Black also – the characteristic of not
quickly publishing in print what they discovered, and of
not even sending it, enclosed in sealed envelopes, to academies
(which proceeding obviously can only have a purely selfish
object); they allowed it to ripen in themselves as far as
possible, without however in the meantime hiding it completely from sight.   It thus came about that each knew more
or less what the others were doing, and at times they worked
simultaneously on the same subjects, so that much appears
as their common achievement.

In another respect also, the three investigators were
similar. The mass of new things they discovered contained
– for them obviously – a great quantity of unknown matter.
It is true that many gases were now known, some of them
combustible (hydrogen), some supporting combustion to
an astonishing degree (oxygen); and they could be weighed
with certainty, just like liquid and solid bodies. But the
phenomena of combustion, and the heat, and also the light,
that were produced, remained something completely
mysterious. The heat of course could even be measured
as regards its quantity by Black's method, like a substance
that could be weighed; but it was not known whether it

weighed anything – anything positive, or even negative. It seemed in some way to be a constituent of the combustible body, a constituent which became free upon burning, perhaps identical with the fire-stuff, phlogiston, which had been imagined for some eighty years as passing from one body to another in the form of flame, when burning takes place. But after the weighings carried out since Black's time, and even by Boyle, difficulties had arisen with respect to weight in the case of heat, phlogiston and light, even if one had been willing to regard the invariability of the weight of all things concerned in the reaction, as assured in the case of these mysterious phenomena of fire. For this it would have been necessary to take up a strongly materialistic attitude, regarding what cannot be weighed as of no importance, and this was just as little characteristic of these three investigators as of Kepler or Newton. This alone perhaps explains the manner in which they all held firmly to the concept of phlogiston,[1] which was even defended energetically by Priestley,[2] but it was just these three investigators, who, as Boyle and Black had already done, did everything towards learning by the only possible means – namely common observation of nature – how matters really stand as regards weighable or unweighable stuff taking part in chemical transformations. With the addition of their investigations, just enough facts were known for an attempt to be made with some certainty to discover whether we may reckon with invariability in the weights of the substances concerned in a

[1] Perhaps this explanation does not hold in Scheele's case.   He says in his *Chemical Essays on Air and Fire* (1777): 'that light must be regarded like heat, as a substance, cannot be doubted.'   According to this, he cannot have been well acquainted with Guericke and Huygens.   Alone the multiplicity of the things considered as of uncertain weight – heat, light, phlogiston, – must have much increased the difficulty of arriving at more correct views.

[2] May it not be that Priestley was also influenced by a justifiable disinclination to accept, before exhaustive proofs had been given, the ideas of a man who, like Lavoisier, had made so many incorrect statements with regard to facts relating to the origin of discoveries?

reaction – taking the gaseous bodies fully into account – ignoring the weight of heat produced or disappearing, and leaving a special fire-stuff (phlogiston) out of consideration. This was first carried out by Lavoisier, who lived from 1743 to 1794 in Paris.[1]

The new discoveries showed that everything was much simpler than people had previously dared to think, and it further continued to appear that in the case of all different kinds of substances, of which new varieties were continually being discovered and in all their changes, combinations, and separations, we only have to deal with matter of invariable weight, which simply arranges itself in various ways. Heat, and light, which for Scheele were still objects for investigation on the same footing as the gases with which he was concerned, thus fell outside the limits of the problem; chemistry became purely the science of the internal composition of the

[1] Lavoisier also lays the main stress upon the facts, only that he allows it to appear as if they had all been discovered by himself, whereas in very many cases he is only giving reports or variations of what had become known to him in other ways from his contemporaries, especially from Black, Priestley, Cavendish, and Scheele, or from what had already been published. A particular mistake often met with is that of imagining the first use of the balance in chemical investigation to go back to Lavoisier. For not only had Black carried out important and illuminating investigations by means of the balance, but even Boyle, and others before him, had already followed the changes in weight taking place in chemical transformations, for instance, the increase in weight when metals are calcined (oxidised). It is obvious that the use of the balance greatly increased as soon as the invariability of the weights of the substances taking part in chemical change had been rendered probable; and that after the discoveries of Scheele and Cavendish had taught us the composition of air and water, and also of nitric acid, the investigation of other substances, such as carbonic acid, sulphuric acid, was soon carried out successfully; this was done by Lavoisier, but he did not state correctly the origin of these discoveries. For details of all this the reader is referred to Kopp's *History of Chemistry* (Brunswick, 1843), vol. 1, pages 302–312, vol. 3, pages 204–206, 266–271. Kopp, who had a thorough knowledge of history, remarked in the latter place quite rightly: ' . . . but the object of history is not to make laudatory speeches, and not its smallest, and for our time, least useful, task, is to show how every appropriation of the achievements of others later becomes revealed, and produces precisely the opposite effect in the case of those who hope thereby to increase their own reputation.' Compare previous note concerning Lavoisier and Black, p. 128.

matter which can be touched and weighed.   The fact that
heat is not matter of this kind was proved a little later by
Count Rumford.

Wilhelm Scheele was born in Stralsund, being the seventh
child of a merchant; in accordance with his wish he became
an apothecary, and lived as such in different localities in
Sweden, finally in Köpingen on the Malasee, where he died
in his forty-fourth year.   At the age of fifteen years, as
apprentice to an apothecary, he already began in his leisure
hours – and when they were not enough, at night – special
studies in all the available writings concerning assaying, and
also experiments on his own account in the laboratory of the
apothecary.   He continued this activity during his whole
life; it was his greatest joy and chief care: 'to explain new
phenomena, that is my care, and how glad is an investigator
when he finds what he has sought so industriously; it is a
pleasure that fills his heart with joy.   For it is only the truth
that we wish to know, and what joy it is to have discovered
it,' he says himself in letters.   Though his means always
remained very modest – it was only in the latter years of his
life that he became the owner of an apothecary's shop, and
he rejected salaried posts which would have tied him – he
replaced the want of them by persistent skill, and a rare eye
for the essential.   A fortunate element in his development
was the fact that his mind had been protected from a long
period of compulsory education at school, and that he had
been spared professional examinations.[1]

He was thus able to absorb freshly and freely what suited
him, without having used up his mind previously on unsuitable material.   He read books through once or twice, never
needing to look at them again; he had obtained what he
wanted.   The enormous number of experiments which he

[1] When he was already known as a learned chemist and trustworthy
man, and took over his own business, the legal examination was turned
into a small festivity.   On this point see: Nördenskjold, *Scheele's Letters
and Laboratory Notes*, with a biography, Stockholm, 1892.

had carried out since early youth, simply in order to see their results, which he then always remembered accurately, had enabled him to collect a store of knowledge and observations of nature, such as none of his trained contemporaries possessed. And since he did not work according to preconceived principles, he saw a great deal and was able to discover a great deal that a systematically trained person would have regarded as impossible, because it conflicted with his principles.

Scheele's discovery of 'fire-air' – oxygen as we call it today, – links up with the fact already known to Leonardo and Guericke, that ordinary air consists of two components, one of which supports combustion and breathing (oxygen), while the other is incapable of doing so (nitrogen). Scheele was successful in separately preparing the first of these. He obtained it from saltpetre, from oxide of mercury, from manganese dioxide (pyrolusite), and other substances, by heating them. In this experiment, as in other preparations of gases, he tied to the neck of the retort which was inserted in the fire, a pig's bladder previously pressed flat, and this then filled with the gas developed. Only when he wished to examine the gas more carefully, did he allow it to rise up under water into a glass vessel. In the latter he was then able to observe the characteristic property of the new kind of gas, the surprisingly brilliant combustion in it of carbon, sulphur, and phosphorus. The burning phosphorus consumed all the 'fire-air,' so that when the experiment was undertaken in a closed vessel, a vacuum remained, whereby a thin-walled vessel was even crushed by the external pressure of the air. Phosphorus, and also a solution of liver of sulphur, which likewise absorbed oxygen completely, gave him the means of determining the amount of oxygen contained in the atmosphere. He also found that nitrogen has a less specific gravity than oxygen, and that the latter is more soluble in water than the former, which must be advantageous for the breathing of marine animals. This work was done in the years 1760 to 1773, but great delay occurred in publication in print, which first took place in 1777.[1]

Scheele not only discovered oxygen from pyrolusite, but also made a very thorough examination of the latter substance. In the course of this he discovered chlorine gas (by bringing pyrolusite in contact with hydrochloric acid) studied the compounds of manganese, and discovered the compounds of barium, which are generally contained in pyrolusite, but had hitherto not been recognised. He showed that they afford the best means for detecting sulphuric acid. How thorough Scheele was in all his work is also shown by his investigation of water. It had been assumed that pure water, when boiled for a long time, turns into a solid substance. Scheele, 'in order to see with his own eyes,' kept distilled snow-water boiling in a glass vessel with a very long neck for twelve days. The water became somewhat milky and formed a white precipitate, but Scheele was able to show that both the precipitate and also the substances found dissolved in the water, were simply constituents of the glass, and that the vessel had become etched upon its internal surface as far as the water reached. No transformation of the water had therefore taken place, and our knowledge of this liquid was thus further advanced.

Scheele was also the first to investigate the various acids of the vegetable kingdom, such as tartaric acid, malic acid, citric acid, oxalic acid; and also uric acid and lactic acid, finding how to separate them by precipitation and transformation. He also discovered glycerine, as a special substance contained in all fats, and separated it from them by means of litharge. He also discovered molybdic and tungstic acids, and investigated them as special

---

[1] The first proved entry by Scheele concerning oxygen in his laboratory notes dates from November 1772 (compare Nördenskjold, *ib.*, pages 462–466). Bunsen described in his lectures the day on which oxygen was discovered as the 'real birthday of chemistry.'

substances, along with a number of others which cannot be given here.

Another achievement of Scheele's was the recognition of heat radiation as a special form of the propagation of heat, different from the well-known conduction of heat, such as takes place in metals, and from the transference of heat by convection, as in liquids, which was already recognised by Black. The fact that concave mirrors and lenses collect the heat of the sun's rays along with its light at their foci, had been known for a very long time; but the heat at the focus was regarded as an effect of visible light, and it had not been recognised that invisible rays also exist which produce warmth, and in other respects also, behave like light. Scheele made this clear by a series of simple observations and experiments upon the fire of a furnace. The heat of the fire is emitted by convection of the heated air in the chimney, and also of that in the room; but it is also emitted in straight lines, as can be proved by using screens, and this takes place without the air through which it passes being warmed. This 'radiant' heat can also be collected by means of a concave mirror, and sulphur can thus be ignited fairly far away from the fire. Here it is not the light of the fire that produces this effect; for the experiment succeeds equally well when the coals are hot but not glowing, and it fails when a glass plate is interposed between the fire and the concave mirror, though this plate allows the light to pass through it. But the plate becomes warm, obviously because, though it does not absorb the light, it does absorb the radiant heat. At a later date (in the year 1800) W. Herschel, the discoverer of the planet Uranus, showed that the invisible heat rays discovered by Scheele fall, when refracted by a prism, outside the visible spectrum, since they are less refracted than red light. Later they were suitably named the infra-red rays.

Scheele also investigated the already known blackening of

silver compound by means of light; he proved that the black is finely divided silver.[1]

Using pure chloride of silver, he observed the blackening of it in the spectrum of the sun produced by a glass prism, and found that it was stronger in the violet than in the other colours. When J. W. Ritter, of Jena, repeated the experiment in 1802, in a somewhat more refined manner, he noticed that the blackening extends considerably beyond the violet end of the spectrum; he thus discovered the invisible ultra-violet rays.

Joseph Priestley was born in a village near Leeds in England; he became a clergyman, and was occupied in turn as a preacher and teacher, but he always busied himself with scientific studies apart from his occupation. His idea of Christianity was essentially different from that general in his country, and this produced many enemies for him.[2]

When in the year 1791, the second anniversary of the outbreak of the French Revolution was celebrated (the deepest causes of which were certainly unknown to Priestley), he gave unchecked expression to his approval, and his enemies seized the opportunity to stir the mob up and destroy his house and laboratory near Birmingham, where he lived at that time, and to plunder it and burn it down. Priestley was scarcely able to save himself and his family, and no longer finding sufficient security in his own country, he emigrated to America, where he died nine years later at the age of seventy-one. In spite of all this, he had faithful friends

---

[1] The art of photography has developed on the basis of these fundamental facts by gradual addition of all sorts of devices, to a continually increasing degree of refinement, without any fundamentally new facts having been discovered. The first fixation of a photographic image produced by short exposure and subsequent development was effected by the painter Daguerre in Paris, in 1835.

[2] He also published in 1782 a work called *History of the Corruptions of Christianity* (he had also lectured upon this theme previously), in which he devoted great pains to arrive at more truth concerning the life and doctrine of the Founder of Christianity, than was usually accepted.

WILHELM SCHEELE

JOSEPH PRIESTLEY

HENRY CAVENDISH

among the best in England, the few who really knew him; among these were Boulton, Watt (who almost suffered the same fate), and other eminent persons in industry, science, and art in and near Birmingham, which together formed the Lunar Society (also known as the Lunatic Society), since it always met at full moon. In this society there was much discussion concerning the contemporary progress of science. Priestley was in many respects very similar to Boyle; a relentless seeker and defender of the truth, satisfied alone by the joy in new discoveries and observations, with a youthful enthusiasm for all that he undertook, and deeply religious.

His own studies of gases began in connection with Black's work on carbonic acid, which he obtained from a brewery. He was the first to introduce the collection of gases over mercury instead of water, which enabled him to recognise and study many gases as peculiar kinds of matter; for these gases, though they must have appeared on occasion before, could not be properly collected and observed. Using our present terms these were nitrous oxide, nitric oxide, carbon monoxide, ammonia, sulphurous acid, gaseous hydrochloric acid, silicon fluoride. He must therefore be called the discoverer of all these. Oxygen and sulphuretted hydrogen had already been discovered by Scheele, but this was no doubt unknown to Priestley, who published the discovery earlier (1774), in his work on the different kinds of air.

Priestley was also the discoverer of the fact that plants consume carbon dioxide and exhale oxygen, whereby the air always remains constant in its composition and suitable for the breathing of animals; he also showed that this phenomenon only takes place under the action of daylight. He also investigated the strength of a sound produced by a bell in different gases, and found it to be very feeble in hydrogen, and distinctly louder in carbon dioxide than in air.

Henry Cavendish belonged to one of the oldest and richest

Ls

families in England ; he lived almost continually in London, in solitude and great retirement, devoted entirely to science.

Cavendish discovered in 1766, that is before Scheele's and Priestley's experiments on gases, hydrogen gas as a peculiar body entirely different from ordinary air. It was the second, after the carbonic acid already fully investigated by Black, of the kinds of gas recognised as fundamentally different from atmospheric air. It had of course long been known that iron dissolves in acids with effervescence, and Boyle already noticed the inflammability of the kind of air which was emitted, but without investigating it further. This investigation was carried out with great thoroughness by Cavendish. He noticed, and also was the first to determine, its strikingly small specific gravity, whereby he also took temperature and pressure into account.[1] He measured the quantities of gas developed by the solution of given weights of iron, zinc, and tin in different acids, investigated the explosive mixtures of hydrogen and air, and made many other contributions to our knowledge of this gas.

He also investigated later the gaseous products of combustion of different substances, and found that carbonic acid is only produced in the combustion of vegetable and animal substances.

In the year 1773 he carried out electrical investigations, which led him to the law of inverse squares for electric force, and even to an acquaintance with what we call to-day the dielectric constant, but he left all this unpublished.[2]

[1] Black soon afterwards pointed out that thin vessels filled with this gas would rise in the air. Seventeen years later, this was put into practice on a large scale by Charles, professor of Physics in Paris, and thus our present-day airships originated.

[2] See Maxwell, *Papers II*, page 612, and *Treatise on Electricity and Magnetism*, 1892, vol. 1, pages 80 ff. These are investigations which were taken up more than ten years later by Coulomb, and in other matters sixty years later by Faraday, and carried further. Scarcely anything else, apart from his published successes, could give us a higher opinion of Cavendish as a scientist, than this pioneer work, combined with his holding back results which obviously did not completely satisfy him.

Very important was Cavendish's discovery in the year 1781, that hydrogen and oxygen when burning together form water, he also showed that the weight of the water formed is equal to the weight of the two kinds of gases which have disappeared. This highly unexpected and astonishing result of the production of water from these newly discovered gases soon became widely known, after Cavendish had first himself communicated it to Priestley; it was the key at last to the composition of water, which had so long been considered, since the ancients, as a simple element. From then on progress was rapidly made in discovering the composition of other bodies, also made up of gases. Cavendish delivered a further important contribution to this, when he showed (1784) that oxygen and nitrogen, mixed together in suitable proportions, disappear when exposed to continual electric sparks, while nitric acid is formed. For this reason, Scheele's fire-air later received the name of oxygen, or 'acid-former,' as being a constituent of acids. It is noteworthy that Cavendish already discovered that a certain proportion of the mixture of air and oxygen remains unchanged even after long passage of the electric sparks. This was the substance only recognised as a separate element a hundred years later, and then called *argon*; a regular constituent of atmospheric air very similar to nitrogen.

An entirely different kind of investigation of the greatest importance, and one which could only be carried out with the greatest possible skill in the art of quantitative observation, occupied Cavendish's later years, up till 1798; this was the measurement of the earth's density. Stated more correctly, it was not only the first proof, but also measurement, of the force of gravity between two terrestrial masses small enough to be contained in an ordinary room. The fact that any two masses exert an attraction upon one another, depending according to Newton's law upon their magnitude and distance apart, was of course, scarcely in doubt after

Newton's researches. But nevertheless, what is required is experimentally proven knowledge, and the proof of the force, which according to the law must be very small, between two not very large masses, and hence had always remained unnoticed, was an important rounding off of our knowledge of gravitation. Conversely, the measurement of such a force in the case of two masses known by weighing, gives us the means of estimating the mass of the earth with accuracy in the same units as that in which the two masses are measured. In Newton's case, on the other hand, the mass of the earth formed the unit in which the masses of the other heavenly bodies were measured; but this unit itself remained unknown, and could only be estimated upon the assumption of a mean density for the earth's material. It now became possible to determine this density with certainty as the quotient of the total mass and total volume of the earth.

In order to measure this very small force to be expected, Cavendish made use of a simple apparatus, the torsion balance, which had shortly before been made use of by Coulomb for another investigation of equal importance. It consisted of a long thin wire, at the lower end of which hung a light horizontal rod, carrying at each end a lead ball. It was then possible to arrange near the two movable lead balls, two much larger fixed balls of lead, which by their attraction upon the smaller ones would produce a deflection of the rod and twist the suspension wire. The measurement of the very small deflection was difficult; the slightest draught of air made it impossible. The apparatus was therefore observed through an opening in the wall by means of a telescope in the next room; and in other respects also, Cavendish showed the greatest care in this, the first measurement ever made of such extraordinarily small forces. His result for the mean density of the earth, namely about 5.5, has been entirely confirmed by much later measurements according to methods in some respects very different. It implies, since the mean density of all substances on the surface of the earth is much smaller, that heavy substances, no doubt metals, must be predominant in the centre of the earth. This is to be expected if the earth was originally a molten mass, since the heavy substances would have collected near its centre, but we thus know for certain from the density as actually measured, that these substances must be present in the interior in considerable amount.

CHARLES AUGUSTIN COULOMB
1736–1806
AND HIS PREDECESSORS

WITH Coulomb, the knowledge of the peculiar electrical and magnetic phenomena, which had been dimly apprehended from early times, begins to develop into a science; for he made the first quantitative determinations of the laws of both phenomena; the two 'Coulomb's laws.' Upon this followed a rapid and quite unexpected development; on the one hand, fragmentary existing knowledge concerning electrical phenomena was expanded into a complete system of 'electrostatics' (theory of stationary electricity), and on the other hand, magnetism was provided with the foundation upon which Gauss and Weber were later able to build. This then led to the quantitative mastery of all electrical and magnetic quantities, which finds its expression in the construction of the system of units generally used to-day, in which – by way of recognition of the historical connection – the technical unit of electrical quantity is called after Coulomb.

Coulomb's law of electricity states the forces with which electric charges attract or repel one another, and tells us that these forces are always proportional to the quantities acting upon one another,[1] and inversely proportional to the square of their distance apart. Coulomb's law for magnetism is also exactly similar. Both laws are thus completely analogous to Newton's law of gravitation discovered one hundred years previously. In spite of their simplicity and similarity with what has long been known, the recognition and indubitable proof of these laws was nevertheless at that time a rare masterpiece of experimental skill, and more than this, it required means to be found by which the peculiar electric and magnetic phenomena could be grasped quantitatively, in order to bring to bear on them Pythagoras' ancient insight – the power of number. The forces to be measured are very small, and they are very fleeting, for electricity itself, as Coulomb himself found, and also observed quantitatively, even dissipates itself into the air.

For measuring small forces, Coulomb worked out the special method of the torsion balance, which has since been made use of in a large number of the finest measuring instruments (galvanometers, electrometers, etc.), all of which depend upon allowing the forces to be measured to twist a very thin fibre or wire, upon which they act through the arm of a lever, suspended in a horizontal position by the fibre or wire. The amount of the resulting angle of twist, at which the force to be measured is in equilibrium with the opposing elastic force of the fibre, is then a measure of the force. Coulomb was thus obliged to examine the laws of the forces required to twist fibres and wires. He found them to be proportional to the angle of twist, and also to the fourth power of the diameter of the wire, inversely proportional to

[1] The proportionality of quantity was not proved by Coulomb by means of special experiments, since he takes the forces from the start as a measure of the quantities of the unknown electricities and magnetisms, but not without having previously proved that this assumption can be carried out consistently, by numerous experiments, for example on the division of quantities of electricity between conductors brought in contact with one another.

its length, and independent of the load on the wire. This
was the result of an investigation published in the year 1784,
and important in itself in regard to the knowledge of the
elastic properties of matter. An interesting point in it was
also the method for measuring the force of which he made
use. He allowed a body suspended by the wire to execute
torsional oscillations about the wire as axis, and measured
the period of oscillation with different moments of inertia of
the suspended body, from which the elastic force governing
the oscillations can be calculated. This was a new application, the first of its kind, of Huygens' and Newton's investigations of the pendulum, to the measurement of forces other
than that of gravity. Coulomb further applied this measurement of force by means of oscillations (dynamic measurement
of force as opposed to static measurement with the torsion
balance) in numerous experiments, directly to the measurement of electric and magnetic forces. The first to follow
him in this method of measuring force and also in the
application of the torsion balance, was Cavendish, who
applied it in his important investigation of gravitation.

Coulomb was also active in other directions. He completely investigated the simple laws of sliding friction, according to which it is proportional to the force with which
the two bodies are pressed together, and independent of the
size of the surface of contact, – as Leonardo had already
recognised – and within certain limits independent of the
speed of motion, but larger as we approach rest. In one of
his last researches (1801), he investigated the internal friction or viscosity of liquids, which Newton had already considered, by means of torsional oscillations, executed by
cylinders hung up in the liquid.

He came of a family of high social position in the south
of France, studied mathematics and science in Paris, and
then entered the army. In accordance with his tendencies
and capabilities, he joined the technical troops. As officer

of these he supervised the building of fortifications during a nine-year stay in Martinique. On his return in 1776, he found for the first time leisure for purely scientific work, and in the thirteen years which followed up to the outbreak of the great revolution, he carried out his fundamental electric and magnetic investigations. They brought him recognition, military advancement, membership of the Academy, and an influential position in the academic world, not without difficulties having occurred now and then; thus it is reported that he was condemned to arrest, when an outspoken technical report of his had made him disliked in high places of the government. When the weakness of the king became greater and more obvious, and the rule of the mob had begun (storming of the Bastille, proclamation of the Rights of Man, 1789), Coulomb resigned all his official posts and retired to his small estate near Blois, where he lived entirely for his science and for his family.[1] Napoleon, who had restored order, gave him back his former posts in which he worked devotedly until his death in Paris at the age of seventy years.

Coulomb's fundamental achievement as regards electricity is best appreciated, when we consider how much detailed knowledge had been collected up to his time, without it being possible to obtain a general and comprehensive idea of it. In this connection, we may recall several eminent investigators, more or less forerunners of Coulomb, who by careful experimenting and clear thinking, had already brought to light from time to time new forms of electrical phenomena. All their work appeared like the 'effects of a magic force,' as Priestley expressed it;[2] as a result of Coulomb's

---

[1] In this respect he behaved quite differently from his contemporary Lavoisier, who adopted the slogan of the revolutionists: 'Il faut tout détruir – oui, tout détruir – parcequ'il faut tout recréer' (Everything must be destroyed – yes, everything destroyed – since all must be created afresh); he himself however, ended by the guillotine.

[2] In his *History of Electricity*, 1767.

work, it was suddenly brought nearer to being understood as a unity, inasmuch as all phenomena could be referred to two electricities and two magnetisms, – themselves not less wonderful, of course – acting according to the law of inverse squares.

The first discovery of magnetic and electrical phenomena is of very ancient origin. The lodestone, or magnetic iron ore (no doubt so called from its place of first discovery, the town of Magnesia in Lydia), with its property of attracting small particles of iron, and the similar property of amber (called by the Greeks *Elektron*) when rubbed, of attracting small bodies, was known long before our era. Artificial magnets of steel were also made by stroking with lodestone, and it was known very early that magnets, suitably arranged so as to be free to move, pointed in the direction of north and south, a matter of great importance for navigation on the open sea. The small deviation from the true meridian, called the declination, had also been noticed. Columbus noticed on his journey to the west that the easterly declination of the Mediterranean gradually reversed in direction and became increasingly westerly, which caused him great concern in view of his complete ignorance of the cause; for when the sky was overcast he had no means of keeping on his course except the compass.

Only in the year 1600 was further progress made. Gilbert, Queen Elizabeth's doctor, made extensive experiments with artificial magnets; he was indeed the first person who investigated the phenomena scientifically. He found among many other things, the fact that could only be fully appreciated after Coulomb's work was known, namely that although the magnetic forces proceed from the poles, of which every magnet has two, and of which those pointing in the same direction repel one another, it was never possible to obtain a single pole separately. For if a lodestone or an artificial magnet is broken into pieces, each piece again exhibits two poles, which lie in the same directions as those of the original magnet. He further recognised, in investigating the poles by the aid of a small suspended needle, that they are never points, but always more or less extended regions of the magnet, from which the attraction or repulsion proceed. These facts, which Coulomb also had to take into account very closely in the course of his investigations, and which altogether pointed to the need for great caution when attempting to calculate with magnetisms or magnetic fluids supposed to be located at definite points in space, and to be the bearers of the magnetic force, could not be explained until more than 300 years later, after Ampère and Faraday had arrived at a better understanding of the matter; their perfectly clear revelation at so early a period by Gilbert is thus all the more admirable. The same is also true of his recognition of the fact that the earth is a great spherical magnet, which at once allows us to form a general, accurate, and comprehensible picture of the behaviour of the magnetic needle, including its inclination (deviation from the horizontal, when hung up at its centre of gravity) at all points of the earth's surface.

In the matter of electrical phenomena, Gilbert's only contribution to progress lay in his finding a number of other substances beside amber – for example precious stones such as diamond and sapphire, glass, sulphur, resin – which are also electrified by friction, and in recognising clearly not only the similarity between the electric and magnetic forces, but also their definite points of difference.

Seventy years later followed Guericke's first step towards the construction of an electrical machine, Leibniz's observation of electric sparks produced by it, Guericke's discovery that electrical repulsion also exists as well as attraction, and Boyle's proof that electric and magnetic force also acts in the vacuum of the air-pump. Thenceforward, thunderstorms were regarded with increasing certainty as electrical phenomena.

Sixty years later again, Stephen Gray in London found that electricity can be conducted; that there are conductors and non-conductors of it. He carried out the conduction of it over increasingly greater distances; in the year 1729, by means of a string of hemp several hundred feet long and supported by silk thread, he conducted it from a glass rod electrified by friction to an ivory ball, at which the phenomena of attraction could then be observed. From that time it was possible to speak of the 'flow of electricity along conductors,' and it was now more than ever described as a 'fluid.' Gray also was the first to set things, and also persons, who were to be 'charged' with electricity, upon an 'insulating stool,' a cake of resin, and he recognised that a hollow charged cube of wood behaves exactly like a solid one of like size; from which it must be concluded, that electricity has its seat only upon the surface, and not in the whole interior of a body. Very noteworthy is also his observation of the non-interference between electrical and magnetic forces; a key held up by a magnet could be insulated and electrified, and attracted light test bodies, in just the same way as in absence of the magnet.

Not much later, in 1733, Dufay in Paris made the highly important discovery that there are two kinds of electricity, which behave in opposite ways, and were therefore called positive and negative. He found that gold-leaf electrified by glass, and hanging in the air, was not repelled, but attracted, by electrified resin, and he followed this up closely; but it was some time before his discovery was recognised.

Ten years later a new observation was made. As the electrical machine was gradually improved, it was fairly natural for the attempt to be made to fill insulated bottles with electricity, which was now available in plenty. Ewald Jürgen von Kleist, dean of a cathedral in Pomerania, and later president of the High Court of Justice (and of the same family as the poet Kleist) electrified a nail inserted in a medicine glass (no doubt somewhat moist inside) and received a sharp shock, when he touched the nail with one hand while holding the glass in the other. The effect was further increased when the glass was filled with alcohol or mercury. These observations made in the year 1745 were soon made known in various directions, and produced great astonishment. A year later the same experiment was performed at Leyden in Holland, and hence the 'intensifying jar' with which numbers of people now began to experiment, was called the Leyden jar; to be historically correct it should be called the Pomeranian or Kleist jar.[1] It was soon given metallic coatings, and great attention was paid to the shocks which it gave, and to its spark effects. Benjamin Franklin showed in 1747, that the two coatings carry opposite electricities. He also remarked, when he made the coatings removable, that the bare glass surfaces are themselves oppositely electrified. This effect of the glass, which goes beyond the part which it plays as insulator (it is called to-day the residual charge, and is not now of great importance), was at that time regarded as of essential interest, until Wilke and Aepinus, two citizens of Mecklenburg, showed in the year 1762, that the intensifying effect is also present in absence of any glass. They covered two large boards with metal foil, hung them up parallel to one another, and very close together, insulated them, charged one positively while the other was connected to earth, and found that the two proved to be oppositely charged, and when touched simultaneously with the two hands, gave a shock like that from a Leyden jar. They were thus able to clear up to a certain extent the phenomenon called to-day electrification by influence or induction, and Wilke had already made experiments with a rubbed plate of glass and a metal plate, which quite corresponded with the later 'electrophorus' of Volta,

[1] In old writings the name Kleist jar is actually found, and might be reintroduced.

and proved the possibility of obtaining by means of influence any required intensification of an existing charge, as happens in the 'influence machine' to-day.

Nevertheless, the idea of the action of the two electricities upon one another, and upon like charges, was still very in-definite, as was also the part played by the matter composing the charged or uncharged bodies in regard to the effect pro-duced, particularly if it could be shown that only one kind of electricity (positive) exists, the other (negative) being only a deficiency of the former, a possibility which could not be rejected. People spoke of an 'electric atmosphere' which every charged body spreads around itself, or of a 'sphere of electrical action.' As compared with this, a great simplifica-tion, and translation from the undetermined to the entirely determined, occurred when Coulomb was able to announce his laws after this state of uncertainty had existed for about twenty years. For these laws actually proved sufficient to render possible a quantitative grasp of all phenomena of stationary electricity (electrostatics), that is of electricity which had reached equilibrium after flowing along con-ductors. Even Faraday's discovery fifty years later of a true and essential influence exerted by the insulators surrounding the conductors, only required the addition of a material con-stant (the dielectric constant) to Coulomb's law, without any change in the form of the latter. Coulomb had himself also investigated very extensively a series of phenomena of sta-tionary electricity, including the distribution of electricity on conductors of different shape (particularly rows of spheres and cylinders), whereby he made use of a 'proof plane' for the first time, along with the torsion balance as a measuring instrument; the proof plane being used to lift off, as it were, the electricity situated at a certain point, on a conductor, and enabling it to be measured. Above all, Coulomb con-firmed by very refined methods the fact already noticed by Gray, that electricity is only situated on the external surface of conductors; and he observed that this also is a consequence of the inverse square law, and can only be true if the latter holds exactly. Every other law would lead to the existence of forces in the interior of conductors, which would set the electricity in motion and change the distribution, a fact which already follows from Newton's consideration of gravitation. He also shows the connection between the distribution of electricity upon, and the curvature of, the surface of a conductor, and the resulting phenomena exhibited by sharp points.

It should be noted that Coulomb had a forerunner in Cavendish as regards the recognition of the fact that the inverse square law of electric force can be deduced from the confinement of electricity to the surface of bodies; Cavendish's observation and considerations on this point, which dated from 1773, had however remained unpublished, and were only discovered much later among his posthumous papers.[1]

LUIGI GALVANI (*1737–1789*)
ALLESSANDRO VOLTA (*1745–1827*)

No sooner had Coulomb rendered the phenomena of stationary electricity comprehensible in a manner for the moment sufficiently satisfactory, than the entirely unexpected discoveries of Galvani and Volta opened up quite new prospects. A fresh source of electricity had been found, which supplied a much more plentiful flow than those hitherto known, friction and influence. Volta's pile took the place of the electrical machine. A new appliance for obtaining hitherto undreamed of knowledge was given us; it became possible to study also the phenomena of flowing

[1] See p. 146 above.

electricity, since the means were given of producing a continuous flow of large quantities of electricity, whereas hitherto only minute amounts, or somewhat larger amounts stored in Leyden jars, had been momentarily set in motion, so that the processes connected therewith had for the most part remained hidden.   In the discovery and utilisation of these processes consists practically everything which makes the 'age of electricity.'   All this only became possible through the work of Galvani and Volta.

The beginnings, as is always the case with great advances into the completely unknown, lay in a modest and devoted effort to understand processes of nature which had hitherto been little or only superficially noticed, which certainly seemed mysterious, but could not be successfully elucidated by already known methods.   Only the rare spirit of born investigators of nature could feel drawn to such activity.   It did not promise external success, not even in the academic world; for the object of the investigation was not even regarded as of any importance.

Galvani was born at Bologna, and began by studying theology, but soon decided upon medicine.   He married early the daughter of his guardian and teacher, worked on the kidneys and ears of fowls, and then lectured from 1762 onwards on medicine at the University of Bologna, where he became professor of anatomy in 1775, and later also professor of obstetrics.

He had quite early made experiments on the excitation of the motor nerves of frog preparations, as we see from a lecture given by him in 1773; but this had to do with mechanical stimulation only.   It was an obvious step to study the electrical stimulation of the frogs; the contraction of living muscle produced by electric shocks was known since the time of Guericke and Leibniz.   During work with the electrical machine in the presence of several persons, it happened in the year 1780 that an observation was made which at once attracted attention as being very curious, namely the fact that strong contractions of the prepared frog legs took place, as they lay upon the table unconnected with the electrical machine and some distance away from it, when the nerves were touched quite lightly with the point of the scalpel. It quickly appeared that this only took place when at the same time sparks were passing in the electrical machine, and further, when the scalpel was held not by its insulating handle, but by the conducting metal part of its blade. This quite mysterious phenomenon was made by Galvani, 'animated by an incredible enthusiasm and desire,'[1] the subject of an extensive and very laborious investigation lasting eleven years, and including many hundreds of experiments on animal preparations, mainly frogs' legs; in the course of this investigation he made, step by step, continually new and further observations. The publication took place in 1791 under the title of 'Essay on the Force of Electricity in the Motion of Muscles,' which is in four parts.

The first part starts from the observation already referred to, and elucidates, by all kinds of changes in the experiment, the conditions under which the contraction of frogs' legs without contact with the source of electricity is most successfully obtained. We can say to-day that Galvani showed the greatest experimental skill in probing depths which could not at that time be further illuminated, although later observers might well have linked up with his work — which did not as a matter of fact happen. It cannot be denied that Galvani was here dealing with electric oscillations produced by the sparks, with electro-magnetic induction, and even with electric waves[2] – all processes which were discovered in another way, starting from Galvani's third part, by Volta, Oersted, and Faraday, but not until a century later, and then

[1] These are his words in the publication.
[2] A very remarkable observation in this category, not upon frogs' legs, is found in the fourth part of Galvani's Essay.

LUIGI GALVANI

CHARLES AUGUSTIN COULOMB

ALLESSANDRO VOLTA

without reference to the observations contained in the first part.

In the second part, Galvani investigates the action of atmospheric electricity on animal preparations. In these experiments, which were carried out in the open air, he found very powerful muscular contractions taking place at every flash of lightning. Here, as in the first part, the frog preparation only acted as a highly sensitive detector of electric forces. This part does not bring anything new, in so far as the electrical charges of thunderclouds had at that time been already sufficiently proved by the bringing down of sparks (as in Franklin's experiments with kites in 1652); this part is only short.

The position is different as regards the third part. This starts from an entirely new observation. For the experiments of the first and second part it had always been necessary to attach conductors – wires or other metal parts – to both the nerves and the muscles of the frog preparation, and Galvani noticed in experiments carried out in the open air, that contractions occurred even under a clear sky, when metallic parts of different nature touched one another; when, for example, the nerve of the preparation was attached by means of a brass hook to an iron garden fence, and the muscle touched the iron. Galvani thought at first that these contractions were to be referred to changes in atmospheric electricity, 'but when I then carried the animal into a closed room, laid it on an iron plate, and began to press upon the latter the hook attached to the spinal cord, behold, the same contractions, the same motions.' Thus a new discovery was made. It again challenged Galvani to an extraordinary number of fresh experiments, whereby he changed the conditions in all sorts of ways, in order to discover the essential point, but every time entirely without the assistance of any hitherto known source of electricity, for it was obvious that a new source of electricity had been found, the seat of

Ms

which remained to be discovered. His chief discovery was the fact that the strength of the contractions depended upon the choice of the two metals, which must touch one another, while one of them also touches the nerve, and the other the muscle; but he also found that the contraction became weaker and somewhat uncertain, but did not quite vanish, when a bow of one single uniform metal was interposed between nerve and muscle. This latter observation, which he made frequently, and always with the same result, led him finally to regard the animal preparation itself as the source of the electricity which caused the contraction. Nerve and muscle together thus appeared like a Leyden jar able to charge itself, which was discharged when its two coatings were connected together. 'Animal electricity' had been discovered.

In the fourth part, entitled 'Some Suppositions and Conclusions,' Galvani attempts to offer some hints, particularly for medicine, as the result of his work. He is here very modest in his statement: 'But all this I have thought out, as I have said, in order that it may be considered by the great and learned . . . in order that they may some day make use of it; that was our main desire' – as he also says in the introduction to the whole: 'For eminent men of learning will be put in a position through reading this essay to develop these results themselves further by consideration and experiment and above all to reach that goal towards which we have striven, but from which we are perhaps still very far.'

Galvani's wish soon reached fulfilment after the appearance of his essay in one direction at least, and in a peculiar way, since Volta took up the study of frog preparations just as Galvani had prepared the way so extensively. He started mainly from the observation of Galvani, that the contraction was more lively when two suitable and different metals were used in the circuit. The endless riches of nature presented him with an entirely new and unexpected discovery; that the point of contact of the different, non-living conductors of electricity act as sources of electricity in just the same way – by mutual production of charge – as Galvani had imagined to be the case with nerves and muscles.   He was also able to throw some light upon the relationship between nerve and muscle.   Nevertheless, Galvani's conclusion regarding animal electricity is also justified, as cannot be doubted in the least according to our present knowledge – although the relationship between electricity and life is hardly any clearer to-day than in Galvani's time.

Galvani was fifty-four years of age at the time of the publication of this essay.   He also took part later in the discussion; but his later years were disturbed by great events. Napoleon Bonaparte had entered north Italy in 1796 as victor, and founded the Cisalpine Republic, to which also Bologna, Galvani's place of residence – hitherto in the Papal States – was to belong. Galvani refused to take the oath of allegiance to the new constitution, and was then declared removed from all his offices.   Though sensible people were able to bring about his reappointment, which was then arranged for the commencement of 1799, he died just before this, in December 1798, at the age of sixty-one.

Volta was born in Como, of an highly esteemed family, became teacher of physics at the High School of his native town at the age of twenty-nine, and was called in 1779 to the University of Pavia.   He travelled much at this time, first in Switzerland, where he became acquainted with Voltaire, who greatly impressed him; later to Paris, where he made friends with Lavoisier and Laplace; to Germany, and to England, where he met Priestley.   Volta, who had a rare gift for experiment, undoubtedly began very early to acquire a thorough knowledge and a profound personal understanding of everything known at the time concerning electrical phenomena.   Very famous for a long time was his 'electrometer,' consisting of two straws, but this was then superseded by the gold-leaf, and later by the aluminium-leaf electroscope; he used it to make measurements of astonishing sensitivity, especially when he added to it the 'condenser,' which consisted of two metallic plates, separated only by a thin coat of lacquer, acting as an insulator, the result being a Leyden jar of high capacity. When the two metal plates were separated, the voltage to be measured on the electrometer was multiplied one hundred times. All this, as well as his electrophorus, was not fundamentally new, but it indicated a development and mastery of the art of electrical measurement which at that time was something quite new, and shared by no one besides Coulomb. Volta also was the first to measure atmospheric electricity, by connecting his electrometer to an insulated flame.

The introduction of measurement, of quantitative observation, into the study of electrical phenomena, of which Coulomb had been the pioneer, proved also in Volta's case to be fertile in results of fundamental importance, when he took up work on frog preparations, in common with many other people, immediately after the appearance of Galvani's paper. In the course of eight years his work resulted in the revolutionary discovery of the 'pile,' and the 'cups,' which were marvellous and completely novel appliances for investigating entirely new regions of knowledge.

First of all, after numerous experiments on frog preparations, which at first were only slight variations on Galvani's, he arrived at the discovery that the muscular contractions are not necessarily conditioned by the transference of electricity from nerve to muscle, but that an electrical stimulus of the nerve alone is sufficient to set the corresponding muscle in motion. He proved this by laying bare a piece of the nerve of the thigh, and providing this with two tinfoil electrodes sufficiently far apart, through which a weak electric discharge could be sent along the nerve without passing

through the muscle. Nevertheless, the muscle contracted
strongly. From then on, Volta used frog preparations as a
test for electrification, exceeding in sensitivity the finest
electroscope provided with his condenser. He now touched
the two nerve points with two different metals, for example
silver and tin, which were themselves also in contact; from
the resulting contraction he concluded that an electric current was passing through the nerve, and says that 'it is clear
that the cause of this current is the metals themselves, and
that the electricity here is excited in a manner of which we
have hitherto had no conception.' He contrasts 'animal
electricity' with 'metallic electricity,' regarding the seat of
the electrification as being the place of contact of the two
metals.[1] In this connection he also made an observation
which had already been known twenty-five years previously,
but which had never been suspected of having any connection with electrical phenomena. If we apply to the tongue
two different pieces of metal, we experience at the moment
when the two pieces of metal touch one another a sharp sense
of a metallic taste. Volta here at once perceived in this the
same effect as with the frogs' nerves, only that in place of the
latter, the tongue forms the moist conductor. He distinguished according to the sour or alkaline taste experienced,
the points of entry and exit of the current on the tongue, and
even used this to determine the direction of the current
given by different metals.[2] Using frog preparations, other
animal preparations, and the tongue, Volta investigated
many different kinds of metal and also carbon and pyrites as
conductors in the solid state, and he already began to arrange
them in a series according to their degree of activity. These
conductors he called conductors of the first class, and liquids,
such as act in animal preparations or on the tongue,

[1] The quotations are taken from Volta's paper published in 1792.
[2] Volta therefore even observed, although without knowing it, the
chemical effects of the electric current.

conductors of the second class, and he laid stress on the continual circulation of electricity through the three conductors, which together form a circuit as long as they are in contact. The seat of the driving force ('electro-motive force' as we say to-day) he regards quite generally as lying in all three places of contact.

Volta now proceeds to eliminate organisms entirely from his experiments. The nerve or the tongue became finally for him only means of testing for electrification; he attempts now to replace them entirely by the electroscope with condenser. The undertaking was very difficult to carry out unexceptionably, on account of the smallness of the forces at work; but in August 1796 Volta was able to announce that he had finally succeeded, beyond his expectations, in making the electrification resulting from the contact of two metals only, 'actually perceptible.' This was the fundamental experiment of Volta, still famous to-day. He succeeded only by using in connection with the electrometers of that period repeated multiplication by means of the condenser, a method in which only a very skilful and careful experimenter can avoid deceiving himself.

Thenceforward the way was somewhat easier. Volta proceeded quantitatively, although only approximately, which however, is generally quite sufficient in the first pioneer work, when the experiments are pure in their results. By investigating pairs of conductors of the first class, he was able to arrange them all in a series, since called the 'Voltaic series,' in such a way that the electromotive force or electrical tension produced by any pair of conductors together, is proportional to their distance apart in the series. At the top of the series stands zinc; at the bottom, carbon; about in the middle, copper; the higher member is always positive towards the lower. This allows us to see that conductors of the first class alone do not permit a continual current of electricity to be produced, since the tensions at the points of contact cancel one another as soon as the circuit is closed. But if a conductor of the second class is included, a steady current is obtained. Volta tested these liquid conductors by soaking porous wood or paper in them, thus obtaining them in the form of flat plates. He found that these conductors on touching suitable metals like zinc, give particularly high tensions, but with other metals, such as copper, only very low ones, so that they do not fit into the voltaic series, and he recognises that this makes it possible, not only to obtain a continual flow of electricity, which had already been assumed in the experiments with frogs and with the tongue, but also by multiplying the points of contact, to multiply the tensions.

With this the voltaic pile was invented. Volta made his first announcement on the 20th March, 1800, in a letter to the president of the Royal Society in London, which was afterwards printed in the transactions of this Society.[1] He there says:

'Yes, the apparatus of which I am telling you, and which will doubtless astonish you, is nothing but a collection of good conductors of different kinds, arranged in a certain manner. 30, 40, 60 pieces, or more, of copper, or better of silver, each laid upon a piece of tin, or, what is much better, zinc, and an equal number of layers of water, or of some other humour which is a better conductor than plain water, such as salt water, lye, &c; or pieces of cardboard, leather, &c. well soaked with these humours; such layers interposed between each couple or combination of different metals, such an alternative succession, and always in the same order, of these three kinds of conductors, that is all that constitutes my new instrument; which imitates, as I have said, the effects of Leyden jars, or of electric batteries, giving the same shocks as they do; which, in truth, remains much below the activity of the said batteries charged to a high degree, as regards

[1] *Phil. Trans.*, 1800, p. 403.   The original letter is in French.

the force and noise of the explosions, the spark, and the distance over which the discharge can take place, &c., only equalling the effects of a battery charged to a very low degree, of a battery having an immense capacity; but which, besides, infinitely surpasses the virtue and power of these same batteries, inasmuch as it does not need, as they do, to be charged beforehand, by means of outside electricity; and inasmuch as it is capable of giving a shock whenever it is touched, however frequently these contacts are made. . . . I am going to give you here a more detailed description of this apparatus, and of some similar arrangements, as well as the most remarkable experiments concerned with them.'

He then also describes the cup apparatus, in which the liquid conductors are placed in cups, each of which contains a strip of copper and zinc not in contact in the cup, but connected outside in the correct order. Each cup thus forms an element of the pile in another form, called to-day the Voltaic element. Volta also describes at the same time some other special observations with the pile, which however all relate to the physiological action of the current: shocks which are particularly strong when both hands are dipped into vessels of water connected to the ends of the pile. Further the continual increasing burning of the skin when the ends of the pile or the chain of cups are touched, which Volta quite rightly regards as a proof of the continual circulation of electricity. He also investigates the effects on the organs of taste, smell, and hearing, and on the eye.

This publication brought Volta an invitation to Paris, to lecture to the Academy, Napoleon Bonaparte also being present; he was at that time First Consul. Much honour was shown to Volta, and this must have given him great pleasure, for he always seems to have been an admirer of France. Five years previously, when Napoleon entered Italy at the head of his army, he was member of a delegation sent to meet the victor. Volta was fifty-five years of age when he discovered the pile; he afterwards made no further contribution to science. He thenceforward devoted himself mainly to his family; indeed, in the year 1804 he also wished to lay down his professorship in Pavia, but this was refused by Napoleon with the words: 'I cannot agree to Volta's resignation. If his activities as professor are too great a burden, they must be limited. He may give even only one lecture a year, but the University of Pavia would be wounded to the heart if I were to allow so famous a name to be struck off the rolls of its members; furthermore, a good general must die upon the field of honour.' He thus remained for a considerable time suitably employed; only in the last eight years of his life did he live in retirement, in his native town of Como, where he died at the age of eighty-two. It is not surprising that Volta did not himself make any further use of his pile, that completely new means of obtaining an ample flow of electricity, in order to investigate the still unknown and special effects of such a flow; the chemical effects, the heat effects, and the magnetic effects, which however are much less obvious. Experimenting with the pile and the ring of cups was taken up on so many sides and with such haste immediately after Volta's announcement, that the discovery of such effects at once occurred in all sorts of places and in a manner which for the moment excluded serious investigation. It was also ridiculously easy to construct the new appliance: a few bits of metal and rag or cups and some salt solution were sufficient. It was a finished gun, loaded with powder and shot, already invented, and easy to fire. It is therefore not worth while to collect the names of those who were the first to publish this or that observation on the effects of the current; but we must consider the work of Davy, Berzelius, Oersted, Faraday, Joule, in chronological order, which work, in the course of the next seven to forty years resulted in the thorough elucidation of the effects of the electric current. It is noteworthy that the mass of new

material, which became accessible through Volta's discovery of 'metallic electricity,' has generally been associated
with the name of Galvani, though of course these discoveries
would not have been made so quickly without his previous
work; we still use to-day the term 'galvanic battery, galvanic
current, galvanometer, galvanisation, galvanic cautery,
galvanism.'[1]

But it was not only in the domain of electricity that in this
period an enormous harvest of research was reaped from the
seed sown by the devoted work of rare minds such as Coulomb, Galvani, and Volta, work which was handed on to
their successors; in other fields as well this was the case, thus
as regards Black, Scheele, Priestley and Cavendish in the domain of chemical processes, gases, and heat.   Further, as
regards light, Huygens' and Newton's results were still open
to further investigation.   Hence from this point onwards,
there is a particularly rapid succession of important achievements, not however depending so much upon work from
the ground upwards, and this is distributed among a comparatively large number of investigators.   In the next thirty
years alone, up to Faraday's time, we must name not less than
sixteen scientists whose work was remarkable; in the previous thirty years there were only eight, and among these
the men already discussed.

COUNT RUMFORD
1753–1814

WE now come to a mind especially lightly weighted with the
ballast of learning, and thus left free to soar to great
heights, but nevertheless capable of getting at the bottom of

[1] Compare in this connection the later note on *Galvanometer* (under
Ampère, p. 227).

everything that he undertook, and always hitting the essence of the matter without pedantry.   Among the numerous and manifold activities of his varied life, only one, relating to science, concerns us here, namely his endeavours to discover the nature of heat; his other services to science can only be mentioned shortly, or simply referred to.

The famous experiment carried out by Count Rumford in the munition workshop at Munich[1] started from the existing knowledge of heat, according to which it appeared as an extremely mysterious something, the amount of which, it is true, could be measured with certainty since Black's time; the addition of it in many chemical transformations, and also in the processes of fusion and evaporation of substances, was known to be necessary, but on the other hand it failed to affect the balance.   It was thus possible to imagine it as a weightless substance – this was the generally accepted idea, and Lavoisier had introduced the term 'caloric' for this substance – and hence of the same nature as the other forms of matter which chemistry had discovered in ever increasing variety.   They all had the feature in common of being fundamental substances, which proved to be unchanged in amount, although they might appear to vanish, together with their properties, in the course of chemical combinations.   In like manner, heat appears to vanish when ice is turned into water, but reappears again when the water freezes to ice, while certain chemical processes, such as combustion, do not require heat, but produce it.   But this idea of heat necessarily led to difficulties in the case of a common phenomenon, namely the development of heat by friction.   For here heat appears without our being able to find out whence it has come, or where afterwards a deficiency in it is discoverable.   This fact struck Count Rumford when he noticed, in the boring of

[1] Published in 1798 in the *Philosophical Transactions of the Royal Society.*

cannon, how hot the casting on the machine became when the borer was blunt. He therefore undertook a series of experiments, in order to test the assumptions common at the time concerning the origin of frictional heat. Heavy friction usually results in the production by wear of particles or powder. If the particles had a smaller heat capacity than the original substance, of course to a sufficient degree, the liberation of heat in the formation of the particles would be explained. Rumford therefore measured in a mixture calorimeter the capacity for heat of the gunmetal turnings, but found it in no way sensibly different from that of larger pieces of the same metal. It might also be thought that the heat comes from the surrounding air, which has free access. But when borings were undertaken in a closed space, and finally even under water, the development of heat, which was carefully measured in each case, was in no way diminished. In one experiment, boring was then undertaken under water, with an intentionally blunted borer, for an uninterrupted period of two and a half hours; the water in which the metal was immersed came to boiling point, and continued to boil as long as boring took place. 'It would be difficult to describe the surprise and astonishment of the spectators,' says Rumford, 'when they saw so great a quantity of cold water heated up and finally caused to boil, entirely without fire.' He then calculated the quantity of heat produced in unit time, and found it – not including the obvious loss of heat to the surroundings – at least equal to that given by the combustion of nine large wax candles. The borer was kept in motion by two horses, but one only would have been sufficient. This unusual way of obtaining heat by the work of a horse is not however very advantageous, as Rumford remarked; for more heat would be produced by using the horse's food, which cannot be dispensed with, as fuel.

In conclusion Rumford once more raises the question:

'What is heat ? ' and answers in accordance with his observations: 'It cannot be a material substance.'   For it came out of
a limited system of bodies, namely the piece of metal in the
trough of water, continuously, and so in unlimited amount;
none flowed in, yet it flowed away on all sides, without the slightest sign of chemical change appearing, say
in the water.   The piece of metal subjected to friction was
fully demonstrated to be an inexhaustible source of heat:
'It appears difficult, if not quite impossible for me,' then says
Rumford, 'to imagine heat to be anything else than that
which in this experiment was supplied continuously to the
piece of metal, as the heat appeared, namely motion.' This
conclusion was a new legacy to natural science.   In view of
the mass of new material already accumulated, and easier to
deal with, it remained for more than forty years untouched
until a quite original mind again appeared as a true seeker
after truth, and attacked the matter anew: Julius Robert
Mayer.   'Heat as a mode of motion,'[1] could, it is true, not
be announced upon a perfectly sure basis until sixty years
after Rumford's time, when the work of Clausius, and also
the earlier work of Carnot, had appeared.   The fact that in
all the years after Rumford's experiments, his idea, founded
upon the observation of nature, did not prevent the general
retention of the idea of a special heat-stuff, shows how
rarely academic men of learning are seekers of the truth.
Only a few noticed the contradictions to reality in which they
were involved.   Among the few searching and original
minds, we have Davy, who, in order to be sure of the matter,
caused ice to melt by means of friction, and this succeeded
although the capacity for heat of the resulting water was
almost twice as great as that of the ice.   He also caused
wax to melt by friction produced by a clockwork in an

---

[1] The title of a collection of lectures addressed both to scientists and
lovers of nature, given by John Tyndall in 1863 at the Royal Institution
in London.

evacuated space, taking precautions which excluded all possibility of conduction of heat.[1]

Rumford himself carried out further very refined experiments in order to determine a possible weight of heat-stuff. He had already found earlier (1785), that a heated body did not sensibly weigh more than a cold one, and also not less – as some people believed themselves to have proved – if only care were taken to avoid disturbance by currents of air, which a hot body when not enclosed always produces around itself. 'For a long time I feared to form a decisive opinion on the matter,' he says, 'on account of the great difficulties of carrying out weighings of this kind,' until (in 1799) he had invented a special device for eliminating all temperature differences upon the balance, and other sources of error. The result, again completely negative, he then communicated to the Royal Society in London.[2]

We may here remark by the way, as illustrating the inexhaustible surprises which nature prepares for us, that since the fairly recent experiments of Hasenöhrl (1904), there can no longer be any doubt that heat (like every form of energy) does possess weight, though its extraordinarily small amount makes it quite immeasurable by present methods of weighing.

Rumford, whose original name was Benjamin Thompson, was born in North America, of a family of English origin; he was very poor in his youth, could not gain much schooling, and was entered as an apprentice in a business house at the age of thirteen. But he soon set up as a teacher with the knowledge which he had acquired from time to time, and then married a rich widow. When the American War of

---

[1] These experiments were carried out a year after Rumford's announcement; they are to be found in Davy's first publication. See the *Works of Sir Humphrey Davy*, vol. 2, pages 11 ff.

[2] These communications are to be found in the *Philosophical Transactions*, 1799, page 180. According to the figures he gives there, he was able to show that a kilogram-calorie certainly weighs less than – 0.013 milligrams.

Independence broke out, he took service in the American Army; for he had a passion for a soldier's life. But his aristocratic opinions soon made him suspected in America, so that he was obliged to take flight to an English ship. He never saw his wife again; his daughter, at that time in the cradle, only returned to the house of her father twenty years after. He was then in English service, where he soon became Under-Secretary of State in the Ministry for the Colonies by virtue of his attractive personality and tactful and trustworthy manner. He then began his scientific investigations, which first dealt with munitions of war.[1] Later on he joined the German princes; he made so good an impression on the Kurfürst Carl Theodor that he was taken into the Bavarian service, and after occupying various high offices of state, became Minister for War. In this capacity he also had workshops at his disposal, in which he carried out the remarkable investigation above described. He showed great love for his adopted country, carried out a reconstruction of army organisation, and showed the most active care by means of all sorts of inventions – heating, lighting and cooking apparatus – for the poverty-stricken people. He thus became much beloved in Bavaria,[2] where he was given the title of Count Rumford, which he afterwards always used. However, the death of his patron, Kurfürst Carl Theodor (1799), undermined his position in Bavaria, so he decided to live in Paris, where the first Consul, Napoleon Bonaparte, received him with distinction. He died there at the age of sixty-one.

[1] In London he also founded the Royal Institution, an institution for research and lectures, where Davy, Faraday and Tyndal worked. Their place is occupied to-day by Sir William Bragg.
[2] He laid out the 'English Garden' in Munich. The memorial to him there was put up in 1795.

MARTIN HEINRICH KLAPROTH (*1743–1817*)
JOHN DALTON (*1766–1844*)

GRADUALLY, and particularly by the work of such versatile, learned, and energetic investigators as Scheele and Klaproth, the experience accumulated by chemistry, concerning the composition of the vast number of substances found in nature or prepared experimentally, had arrived at the point at which it was possible to distinguish as their components a not very large number of fundamental substances, out of which perhaps all other substances could be built up, and which themselves were possibly of a simple nature, not susceptible of further decomposition. An important matter was the collection of extensive knowledge concerning the weights of the fundamental components contained in various compounds, or necessary to their formation, and in this connection Klaproth especially did exemplary work.

The assumption of the formation of all kinds of matter from certain fundamental substances discovered by experiment only became possible without inconsistency when Black, Priestley, Cavendish, and again Scheele had shown how seriously it was necessary to take the various kinds of 'air' – gases – which they had been the first to discover, as weighable components; for the fundamental assumption was the invariability in the weight of the components, whether separate or in combination, and in however different a form. Heat on the other hand, fire, (phlogiston) – however important it might be in preparing most substances, and hence itself for long regarded as a necessary component – had gradually been left out of these considerations; for it could not be grasped by the balance. Lavoisier had in any case shown that such a limitation in the number of the weighable fundamental substances made everything appear much simpler than one had hitherto dared to think, and a simplification of this kind has at all times been equivalent to the assurance

COUNT RUMFORD

MARTIN HEINRICH KLAPROTH

JOHN DALTON

that our ideas have approached considerably nearer to actual reality.   Furthermore, Count Rumford had directly proved that there can be no question of heat having weight.[1]

From this point, Dalton again made a great advance.   He saw clearly the great importance of the fixed relationships by weight with which any two components combine to form a new substance: the 'law of constant proportion' as it is called.   To this was added the 'law of multiple proportions,' inasmuch as it was found that in many cases several different relationships by weight of the same components were possible, but that these were then multiples of one another, and also, these multiples were always small whole numbers. Dalton had himself specially examined some examples of this.

Along with these facts, Dalton had in mind the extreme divisibility of all bodies; not only of simple bodies, but also of compound ones, such as water, which are not mixtures capable of variation in composition.   Since division, no matter how far it is carried, leaves their properties unchanged, the smallest parts must still contain the components, for example hydrogen and oxygen, in the correct and fixed relationship by weight in which they are always found in any quantity whatsoever of the same substance.   He concluded from this that such bodies as water (we call them to-day chemical compounds, as contrasted with mixtures) must consist finally of very small parts, which are all exactly alike, and which are themselves again composed of still smaller parts of the given fundamental substances, which latter, however, in spite of their obviously imperceptible minuteness, must still have the quite definite weight corresponding to the

[1] In the matter of heat, Count Rumford and Lavoisier agree.   The difference is however, that the practical imponderability of heat was in Rumford's case a fact of observation proved conscientiously, whereas in Lavoisier's case it was a hypothesis introduced *a priori*; further that Rumford – again as the result of direct experiment – did not regard quantities of heat as invariable in all circumstances, whereas Lavoisier introduced his 'caloric,' which was supposed to be invariable in amount.

composition of the substance, e.g. water. These final particles, from which all matter is built up, are called by Dalton 'atoms.' He speaks not only of atoms of the fundamental substances, but also of atoms of water, for example where we to-day speak of molecules, in order to distinguish groups of atoms from single atoms. Such an atom (molecule) of water might in the simplest case consist of one atom of hydrogen and one atom of oxygen (this is Dalton's assumption), and in this case the known composition of water would already give us the relationship by weight of these two atoms (very nearly 1 : 8 according to our present knowledge, 1 : 7 by the figures known to Dalton). But instead of one atom, there might also be two, three, and so on, of each fundamental substance in each molecule; any multiples of whole numbers, but no intermediate stages, since the atoms are supposed to be alike and unchangeable; this gives the explanation of the law of multiple proportions. The possibility of multiples did not prevent the weights of the atoms being determined; they only made it more difficult, inasmuch as many compounds of the same fundamental substance had to be investigated, in order to exclude as far as possible errors due to multiples. This is the line taken by Dalton with the existing knowledge of the quantitative composition of various compounds, and he accordingly constructs for the first time a table of atomic weights, in which that of hydrogen is already taken as unity.

This became the foundation of the whole further development of chemistry, and indeed, of the whole further development of all our knowledge of matter. It was at that time only an hypothesis (supposition), but an hypothesis founded on a very great deal of quantitative knowledge, and one also capable of further quantitative tests. It has stood these tests in the fullest possible manner, inasmuch as it has always led – in cases which to-day are innumerable – to correct conclusions. It has thus gradually become a theory, that is to say a well-founded quantitative piece of knowledge based upon experience, which although it goes far beyond what is directly accessible to our senses, nevertheless possesses the same certainty as direct observations. We are in this way as fully convinced of the existence of the atoms with their weights, as we are of the globular form and axial rotation of the earth, of which we are likewise unable to assure ourselves by the direct evidence of our senses. Starting from Dalton's beginnings, we already know to-day very much more about the atoms than their relationship by weight. Imagined suitably enlarged, though still indefinite in certain details, they form for us to-day quite as much a result of our gradually increasing knowledge, as does the earth's sphere, imagined suitably diminished in size as seen from a distance, with which we are familiar. This was the first great example of our penetration into details of the world not directly accessible to our senses, and was given us by Dalton in the year 1808, in his *New System of Chemical Philosophy*, which was based on the sound investigations of his predecessors, in particular Klaproth.[1]

[1] 'Philosophy' of to-day usually goes back to antiquity when seeking to tell us the origin of our ideas of atoms. It does not notice that it is merely tracing the origin of the word, and it confuses arbitrary imaginings of the human mind – many of which appeared early – with ideas of the external world which are strictly co-ordinated with measurable and quantitative experience; it confuses untrammelled poetic imagining with truth laboriously sought and tested. But in going back to antiquity, it might have learned from Pythagoras how we are to avoid falling into such deadly confusion; by measurement. But even without making invariable use of this method – as the scientist does – since it cannot everywhere be applied, it is possible to serve truth, that is to say to follow out the actual course of events, instead of turning fancies into systems, without paying any attention to their failure when confronted with reality; philosophy alone no longer gives heed to this. In this way, being incapable of grasping our advanced science – a matter however rendered much more difficult by the materialistic method of to-day, and not by the extent of our investigations – it has been condemned to an undoubtedly injurious and merely apparent existence, such as it now leads at the universities, being mainly a terror for examination candidates, and serving to superficialise their minds; and many another moral science is in no better case, likewise for want of an unconditional enthusiasm for truth.

Since then, the atomic weight has not only been determined with very greatly increased accuracy through the further refinement of quantitative chemical analysis, but it was soon learned, by the use of determinations of molecular weight and other methods, how unknown multiples could be avoided with complete certainty. A somewhat longer time, about fifty years, was necessary before the movements of the molecules in bodies, particularly in gases (kinetic theory of gases), were understood, and this then led to our learning the actual sizes of the molecules, and also their absolute weight (in ordinary units); and the knowledge in detail of molecules and atoms, which internally are worlds in themselves, is still being rapidly advanced.

Klaproth was born at Wernigerode in the Harz mountains, went to school there, and then became apprenticed, and later assistant, to an apothecary in Quedlinburg; but he did not become acquainted with good textbooks of chemistry until his twenty-third year, in Hanover. This awakened his desire to find a position in which he could learn more; he went to Berlin, where he became an assistant in, and finally manager of, an apothecary's business. In the year 1780 he was able to set up his own laboratory, from which a very large number of exact researches issued. He was the first to introduce the custom of stating, as the result of a quantitative analysis, not corrected values, but the data actually obtained by experiment. The loss or excess which almost always results in an analysis, had hitherto always been corrected by the analyst himself according to his own notions, which were often affected by preconceived ideas; and the result of observation stated was not the actual result of the experiment, but almost always only the conclusion which it was supposed possible, with more or less justification, to draw from the experiment. Klaproth was the first to introduce the custom of not only publishing one's own convictions regarding the composition of a compound, but also of giving full details of

LOUIS JOSEPH GAY-LUSSAC

ALEXANDER VON HUMBOLDT

the investigation. The agreement of the sum of the weights of the components found, with the weight of substance taken for investigation, now gave a measure of the accuracy of the investigation, and of the trustworthiness of the methods used. The progress since made by analytical chemistry, the discovery of many new substances, depends solely upon this manner of carrying out and communicating investigations, inasmuch as too great a departure of the weight of the components from that of the substance taken can only be caused by a defective method or by the neglect of an unknown component still contained in the substance.[1]

Thus Klaproth himself became above all an essential improver of the methods of quantitative chemistry, which enabled him to discover mistakes made by others, and which often brought him in touch with Scheele's endeavours of like nature. He also became the discoverer of a number of hitherto unknown elements; thus in 1782 of tellurium, in 1789 of uranium in pitchblende, and of zirconia as a special substance; in 1793 he undoubtedly separated strontia from baryta; in 1795 he discovered titanium, and in 1803 cerium. His methods of exact research enabled many of his successors in the next fifty years to effect similar advances. In particular, the gradual recognition of the remaining elements, so far as they are not present in all too small amount in terrestrial substances, was realised in this way. An entirely new method of finding the presence of elements present only in traces was first given us in 1860 by Bunsen and Kirchhoff's discovery of spectrum analysis. Klaproth's achievements received as much recognition as was possible in his country. In 1809, at the age of sixty-six, he was made professor of chemistry at the newly founded university of Berlin, after he had already taught chemistry for several years previously at the Artillery College. He died at the age of seventy-four.

[1] For further information regarding Klaproth see *Encyclopaedia Britannica*, Art. Klaproth.

Dalton was born in a village in the north-west of England, where his father was a weaver and also leased a small estate.   He was taught in a neighbouring school, but already became at the age of thirteen a teacher in his birthplace, where he also helped his father on the farm.   From then on he earned his living all his life as teacher in various schools, at the same time always working further on his own education by means of the help of friends.   He soon published some original work on mathematical questions.   At the age of twenty-seven he went to Manchester, which remained his permanent home.   His activity as a teacher provided him with sufficient for his modest needs; he did not found a family.

His experimental investigations were carried out with simple means; they related mostly to gases and vapours.   It was natural, since temperature measurement by means of mercury thermometers had been developed, that the investigation of the behaviour of the air and also of that of water vapour at various temperatures began.   In particular, the relationship between volume and pressure, and the temperature, was examined; but many contradictory statements were in circulation.   Dalton was the first to produce thorough results.   Vapours were investigated in barometer tubes, in which he allowed the liquid which was to form the vapour to rise to the surface of the mercury; the tubes were then exposed to various temperatures.   He found that for every temperature a corresponding definite pressure of the vapour exists, which is quite independent of the quantity of excess liquid, and of the volume of the vapour.   This is the fundamental fact concerning vapours, which determines all matter of evaporation, condensation, and boiling; it was Dalton's original discovery, and he already produced tables of vapour pressure.   He also found that the vapour pressure corresponding to the temperature is also present unchanged, even when at the same time a gas is present in the space.   The pressure of the gas, which is determined by the law of Boyle and Mariotte, and the vapour pressure, which only depends upon the temperature, do not interfere with one another; they simply add together. But if the vapour is not saturated, that is, if no excess of liquid is present, it behaves like a gas and follows Boyle's law.

In this connection, he also investigated thoroughly the change of volume of gases on heating under constant pressure. He saw that the data then available could not be of much value, since they came from unsatisfactory experiments; the gases had been enclosed over water, and hence gas and vapour had been measured together. He investigated gases dried as thoroughly as possible; air, hydrogen, oxygen, carbonic acid, and nitric oxide, and he showed that they all expand by an equal amount for each degree, or from the ice point to the boiling point by 100/265, of the volume at the ice point. This is the law published in the same year (1802), but a little earlier, by Gay-Lussac in Paris, and to-day named almost exclusively after the latter.[1] The co-efficient of expansion of gases as later determined with increasing accuracy, 1/273, is a little smaller than Dalton's value (and also that of Gay-Lussac). This co-efficient of expansion of gases became of especial importance after the insight furnished by the kinetic theory of gases founded by Clausius, since it gives us the absolute zero of temperature, $-273°C$, at which the volume maintained by the motion of the gas molecules, and hence the kinetic energy of these molecules – the true measure of the temperature – becomes zero.

Dalton also used his vapour pressure table for measuring the quantity of moisture in the air, in exactly the same way as is done with the present day dewpoint hygrometer, of which he is therefore the inventor, although it later assumed a more convenient form.

[1] In older publications, for example Carnot's important work on heat engines (1824) it is called the law of Gay-Lussac and Dalton, which is as well justified as the usual description of the pressure law as of Boyle and Mariotte.

We have also to thank Dalton for the discovery that gases become heated when compressed, and cooled when expanded (against external pressure). The heating effect on compression had been observed not long previously in air-guns, and it then soon led to the invention of the pneumatic 'fire syringe': Dalton showed by thorough experiments that the cause of the production of heat is not friction, but a very remarkable and unexpected property of all gases.[1] This later became of the greatest importance, particularly to the work of Carnot and Robert Mayer.

Dalton's services to science were soon recognised by many honours, which however had no influence on his circumstances in life. 'He was one of those genuine, and now so rare, philosophical spirits, who find so magnificent a reward in the discovery of truth, that they have but little interest in the passing signs of human recognition, as compared with it.'[2] In the year 1833, he received a small pension from the king; he died eleven years later, at the age of seventy-eight.

LOUIS JOSEPH GAY-LUSSAC (*1778–1850*)
ALEXANDER VON HUMBOLDT (*1769–1859*)

THESE two belonged together as men of science, not only because they carried out important work together and published it jointly, but also on account of the continuous friendship which united them in life, however great otherwise the difference between them may have been as regards the individual importance of each.

Gay-Lussac was born in a small town in the southern part of France, where his father was a judge, and studied technical science in Paris. He then filled various teaching

---

[1] See E. Mach, *Wärmelehre.*
[2] Hermann Kopp, *History of Chemistry*, vol. i, page 364.

posts and professorships of chemistry and physics, and also other public offices, which brought him great public recognition. He died at the age of seventy-two, in Paris.

Best known of his work is the law named after him, which states the equal expansion by heat of all gases, and which he discovered simultaneously with Dalton, by performing experiments, just as the latter did, for the first time in a proper manner, with a number of the different kinds of gas then known. The fact that the co-efficient of expansion measured by him, as also by Dalton, turned out a little too large (as we know to-day), was obviously caused by the property of glass vessels, then still unknown, of holding water condensed on their surfaces with great tenacity, and it does not in the least diminish the value of the discovery of this peculiarity regarding the gaseous state, namely that all kinds of matter show the same behaviour without distinction.

Gay-Lussac's work, performed partly in collaboration with Humboldt, also related to gases, being an investigation of the relationship by volume in which gases enter into chemical combination with one another. He first, in 1795, investigated the combination of hydrogen and oxygen to form water, in the manner discovered twenty-four years previously by Cavendish, who was then still alive, and whose data as regards volume already showed at least approximately the ratio 2 : 1 for the two gaseous components.[1] Gay-Lussac and Humboldt found that when they allowed the two gases, mixed in correct proportions, to explode by electric spark in Volta's 'eudiometer,' that this simple relationship holds with striking accuracy. After this, Gay-Lussac alone investigated, in 1808, a very considerable number of further processes of gaseous combination, such as the union of hydrochloric acid gas and ammonia gas, of carbon dioxide

---

[1] Even Volta already had stated that hydrogen and oxygen give the best explosive mixture when mixed in this proportion; only no one knew previously to Cavendish's work that the result of the explosion was water.

and ammonia, carbon monoxide and oxygen, and many others, and in every case he found no less simple relationships by volume. He also calculated in part the relationships by volume, which would appear in the gaseous state, from the known relationship by weight and the specific gravities of the gases or vapours. He is thereupon rightly able to announce 'that two different kinds of gas, when they act chemically upon one another, always combine in the most simple relationship by volume,' and also that 'the volume of the gaseous products or combination is likewise always a simple proportionate number.'

This fact, which powerfully supported Dalton's atomic ideas, was one of the strongest bases for the further development of all knowledge of atoms and molecules, and hence of all investigation of the structure of matter, particularly in the gaseous state. It was first of all evident that, in the simplest case, an equal number of atoms or possibly molecules of different gases would be present in the same volume. In this connection, the notions of atom and molecule were not at that time distinctly separated. Amadeo Avogadro (1776–1856), in Turin, was the first who, already in 1811, decided to assume the existence of special molecules, even in the case of the elementary gases, not consisting of one, but, at least in the case of the ordinary gases then known, of two similar atoms combined together, whereas the molecules of chemical compounds consist of unlike atoms combined together. When the conception of molecules had thus been defined, it was possible to assume the existence in all gases of equal numbers of molecules in equal volumes, by which agreement was found to obtain with all experimental data of chemistry known at the time. All this necessarily remained for a long time mere hypothesis (supposition), and it was, quite justifiably, only accepted with great diffidence. But all subsequent experience only served to confirm it; though it was only put beyond doubt almost fifty years later, when the same result was obtained in an equally well supported manner, namely from the kinetic theory of gases of Robet Mayer and Clausius. 'Avogadro's law' of equal numbers of molecules in equal volumes of all gases (at the same pressure and temperature) is one of the best ascertained facts of science. It is the foundation of all determinations of molecular weight and hence of atomic weight.

Gay-Lussac was also the first to investigate quantitatively the observation made shortly before by Dalton, that gases when suddenly compressed become warm, and when expanded are cooled, the latter only happening when expansion takes place against external pressure, and not when they flow into a vacuum.[1]

Alexander von Humboldt, son of Major Georg von Humboldt, who served in the Seven Years War, received together with his brother Wilhelm, who was two years older, a careful education at the family seat at Tegel, near Berlin. Neither of the brothers went to school until they entered the university. After many years of varied study, Alexander set out on extensive travels into the interior of North and South America, and to Asia, concerning which he published a very large work containing many geographical, meteorological, zoological, mechanical, geological, historical, and other results. From the year 1827, yielding to the earnest desire of the King of Prussia, he lived permanently in Berlin. He first gave lectures on physical geography to a very large circle of hearers at the university, and then turned his attention to writing his second great work, the *Kosmos*. The leading idea of this he himself states in a letter as follows: 'I have been seized with the mad idea of representing in a single

[1] The investigation in which Gay-Lussac proved the latter fact in the year 1807, later became of particular importance since Robert Mayer was able to use it as an unexceptionable basis for calculating the mechanical equivalent of heat; it is reprinted in the appendix to Mach's *Wärmelehre*.

work the whole material world, everything that we know to-
day concerning the phenomena of the heavens and of the
earth, from the nebulae to the geography of mosses on
granite rocks; and this is to be a work stimulating by the
liveliness of its language, and also entertaining.   Every great
and important idea, which anywhere shines forth, must here
be given together with the relevant facts.   It must represent
an epoch of the mental development of humanity as regards
its knowledge of nature.'   This work, to the repeated re-
vision of which he devoted the greatest care, first began to
appear in 1845, and found a very large number of readers; it
occupied him until his death, at the great age of nearly ninety.

Humboldt was an investigating mind, though no single
great success stands to his credit as an investigator.   In his
manysidedness he is similar to Leibniz, and also in the peculiar nature of his personality, which enabled him to exert
great influence on his contemporaries, and also upon the
succeeding age.   This gave him a high view of the calling of
a man of science, and also of the moral value of the results of
investigation; in this sense he also states, in his *Kosmos*, the
results of the investigation of nature as for friends of nature,
and not for exploiters. This power, rarely met with, of
emphasising the value of the achievement of past times, is
certainly of greater value than the production of a new discovery, if the latter does not meet with people who are educated to a proper understanding of it.

HUMPHREY DAVY (*1778–1829*)
JACOB BERZELIUS (*1779–1848*)

THESE are the two great men of science who were the first
to use the electric current, so freely produced by Volta's
pile, for thorough and fundamental chemical investigation;

HUMPHRY DAVY

JACOB BERZELIUS

but in other respects also they belong to the most eminent minds of their time. The fact that the electric current, when led through aqueous liquids, produces decomposition and develops the two gaseous components of water, hydrogen and oxygen, though in itself highly novel and wonderful, could not fail to be discovered as soon as anyone had set up the pile, and dipped its two end wires into ordinary water. It thus soon became known that an appliance of undreamed-of power in chemical investigation was here available. Not long previously, the decomposition of water in red-hot iron tubes had been studied, and it was found that a very high temperature was necessary for this, whereas by means of Volta's pile, the decomposition took place straight away, as it were by itself, and the two components were cleanly separated. But the easier these new effects were to observe, the more unsatisfactory were the experiments which were now quickly published in large numbers. Water supposed to be pure was said to have been decomposed by the current into acids and bases of various kinds, which were indicated at the two end wires by litmus, and a maze of contradictions came into existence. Davy was the first to clear the matter up by a large number of experiments designed to give clear results, and Berzelius soon followed him in the matter.

They showed that the acids and bases only arise from impurities in the water, and that electrical decomposition does not produce any new components, but only the components of water already known, hydrogen and oxygen. An astonishing fact, and one most misleading, was that two quantities of water could be taken in separate vessels, in each of which one of the wires from the pile dipped, the two being only connected through a moist wick or the like, and that then one lot of water only gave hydrogen, and the other only oxygen. But neither Davy nor Berzelius allowed themselves to be daunted by the difficulty of understanding this phenomenon, which as a matter of fact was only thoroughly explained seventy years later by the investigations of Hittorf. They studied thoroughly and persistently, in well-designed experiments, the action of the current on a very large number of substances, particularly with the object of discovering whether they could be decomposed or not.

Here Davy, an investigator possessed of the liveliest mind and gifted with the surest insight into the secrets of nature, made a wonderful discovery (1807). He separated the alkali metals potassium and sodium, the existence of which, though they had never been seen, had already been suspected for twenty years as constitutents of the well-known caustic alkalis. He found them to have very strange and unheard-of properties. He did not succeed in separating them from solutions of the alkalis; only the water was decomposed. He then melted the solid alkalis, free from water, over a blow-pipe, and passed the current through the molten substance. The negative wire then showed curious flame effects, the origin of which could not be satisfactorily followed; but when he connected the ends of a voltaic battery of many plates, which he had specially constructed for these researches, with a piece of moistened caustic potash, which was melted by the current, he saw brilliant globules with a perfect metallic lustre appear on the negative wire; they were the new metal which he was seeking. When immersed in water, they took fire and burnt instantaneously, swimming about on the surface, and caustic potash was again formed. The remarkable nature of the discovery lay not only in the highly unusual properties of the metal discovered, in the knowledge of an entirely new kind of matter; it was now clear that the other and still undecomposed alkalis and earths must also be oxygen compounds of unknown metals. Thereupon, Davy himself soon actually separated sodium, calcium, strontium, barium, magnesium, in the metallic state, and these bodies themselves soon served as further important aids to chemical investigation.

Davy already made a thorough study of their properties in the course of many careful experiments, as far as the means available at the time allowed. It happened, as may so often be observed in the history of science and technology, that the proof of the presence and properties of newly recognised things was sufficient to result very soon in their production on almost any desired scale. A goal, the nature and direction of which can already be seen, is soon attainable in other ways – it no longer needs a voyage of discovery to reach it, but merely a journey. Davy's discoveries thus immediately led to fresh exertions to separate the new metals, if possible, also by purely chemical means, which were found in the case of sodium within a year; by reduction at high temperatures with carbon and iron it was soon produced on a large scale.

In other respects also, Davy's scientific activity was always directed with the highest and most penetrating clarity towards the discovery of fundamental facts. When he first produced the phenomenon of the electric arc between two carbon rods by means of his large voltaic battery, the great development of heat became to him of the deepest importance. He recognised that here, just as in Count Rumford's experiment with the cannon, and in his own experiments with friction, the notion, still at that time current, of a peculiar heat substance is insufficient as an explanation; for no statement could be made as to the origin of the heat of the electric arc, or of that produced by the heating of wires traversed by an electric current.

We must also mention much more remarkable investigations made by Davy. We can only refer here to that concerning chlorine, the elementary nature of which he was the first to demonstrate, after it had been suspected of containing oxygen. He then also proposed the name chlorine for this element, in place of the term hitherto used 'dephlogisticated spirit of salt' (Scheele's term; which, since hydrogen had been discovered, amounted in meaning to spirit of salt freed from hydrogen) or 'oxidised spirits of salt' (Lavoisier). He also discovered and studied many new compounds of chlorine, among these the very explosive chloride of nitrogen. Many other achievements, among them inventions of value to humanity, but which he refused to patent, cannot be mentioned here.

Davy came from a Norman family,[1] which had settled in the south-west of England. His father was a poor wood-carver. He went to school until his fifteenth year, but unwillingly; he wrote later: 'To learn in a natural manner is true pleasure; how evil it is that in most schools it only causes misery.' He then was apprenticed to a surgeon and apothecary. When there he studied all kinds of books very extensively in his spare time. He first paid closer attention to geometry and mathematics, and only later to chemistry, but this then occupied his entire attention so that he soon began to develop ideas of his own. He was also fond of writing poetry. The patients whom he had to bandage liked him greatly, and altogether he was all his life praised for his kindness, readiness to help, and general amiability. At the age of nineteen, he entered an institution which had just been founded with the object of making use for medical purposes of the various varieties of gas then recently discovered. Nothing could have been more advantageous than this position for Davy's activity, for it placed at his disposal a special laboratory for chemical experiment. He devoted special attention to nitrous oxide, which, as well as other gases, he tested thoroughly by breathing it himself, after finding a suitable means of preparing it. He then published in the year 1800 a special work on the subject, which

[1] See his brother John Davy's book, *Memoirs of Sir Humphrey Davy*, London, 1836; also *The Scientific Achievements of Sir Humphrey Davy*, by J. C. Gregory, London, 1930. Davy was of medium height, had light brown hair and eyes, his head was small at the back, his hair of very fine texture and slightly curly; he had an aquiline nose.

Thomas Young

JOSEF FRAUNHOFER

AUGUSTIN FRESNEL

resulted in this gas, called laughing gas on account of its effects discovered by Davy, soon becoming a favourite anæsthetic for small surgical operations. In 1801 he already, as a result of this, received from Count Rumford an invitation to come to London to his newly founded Royal Institution, where he received a still better equipped laboratory, and undertook the duty of giving every year a certain number of lectures to large audiences.[1]

Here he carried out the work with Volta's battery to which we have referred above, and which then also became the subject of his lectures. It may appear astonishing that he was able at once to appear with great success as a lecturer although he had hitherto had no practice of any importance. However, not only did his poetic gifts and his unlimited enthusiasm for his chosen subject ensure this success for him; but he was aided by the striking exhibition of entirely new effects, such as the Voltaic battery and its action, the peculiarities of the alkali metals produced by it, and the electric arc. Since Galileo had been able to exhibit the moons of Jupiter, or Guericke the Magdeburg hemispheres, there had certainly been no such great public impression produced by purely scientific successes. But Davy, by the depth of his attitude towards scientific research, which he here exhibited, and by suitable historical disquisitions, made his lectures a striking and noble means of culture for old and young and for all circles. He thereby firmly founded the reputation of the Royal Institution exactly in the sense desired by Rumford.

This tireless activity brought him high recognition; in the year 1812 he was knighted, and he also received many distinctions from abroad. He married in the same year. His wife was very well-to-do, and this enabled him to make long travels, extending over years, through almost the whole of

---

[1] For a short time before Davy, Thomas Young had occupied this position.

Os

Europe. On these occasions he always took with him a small travelling laboratory, and ideas for work never forsook him. In the year 1820 he was chosen President of the Royal Society, which he regarded as the highest possible distinction, since Newton had once filled this position. After about the year 1826 his powers began to decline; he again sought for recovery in long voyages which were by preference to Italy, the Tyrol, and Switzerland; fishing in the Alpine streams was one of his favourite amusements. But recovery was not to be his lot; he died on his way home, in Geneva, only fifty-one years of age.

Berzelius was born in East Gothland, where his father was head of the school. He studied medicine and chemistry in Upsala, obtained his medical doctorate in 1802, with a work on the effects of Volta's pile on organic substances, and then became a doctor and, later, teacher of chemistry. His work, like that of Davy, was by no means confined to the chemical effects of the voltaic battery, but covered the whole of chemistry. But Berzelius was more inclined to exact quantitative work, and was thus allied, say, to Klaproth. Thus Berzelius was the first to develop methods of obtaining the quantitative composition of vegetable substances consisting of carbon, hydrogen, and oxygen, following paths opened up by Lavoisier and Gay-Lussac. Never since the time of Scheele, surely, had a man of science been in possession of such a store of knowledge as was Berzelius. Since 1807, in his capacity of professor of chemistry in Stockholm, he had trained a large number of pupils, who continued to advance chemistry after his manner.[1]

He perfected Dalton's tables of atomic weight with continually increasing accuracy, and he invented the method, quite general to-day, of using letters as symbols for atoms,

---

[1] Davy had one pupil only, who however was a host in himself: Faraday. It was later said that Faraday was Davy's greatest discovery.

and placing them together, with the suffix giving the number of atoms, to show the molecular formula. Dalton showed atoms as small circles with distinctive symbols written in them; Berzelius' symbolism gradually became more suitable, since the number of elements ascertained to exist increased, whereas for a long time nothing whatever could be ascertained concerning the grouping in space of atoms in molecules, so that a scientist desirous of adhering to truth, such as Berzelius, felt obliged to avoid for preference the suggestion of grouping in molecular formulae.

Berzelius received the highest recognition and distinction. He married in 1835. He lived to the age of sixty-nine. Both Davy and Berzelius – the first somewhat earlier[1] – made use of their experiments on the chemical effects of the electric current to draw conclusions concerning the nature of the atoms; they recognised that the atoms either contain electric charges from the beginning (Berzelius), or must be capable of communicating them to one another by contact, either singly in the molecule, or in large masses in the case of the plates of the Voltaic battery (Davy). This was the first time that properties of the atom appeared to be capable of deduction on the basis of well-founded experience, and these properties were electrical in nature. A further eighty years were to pass until – by the study of Hittorf's cathode rays – new experience of an essential description was again obtained, but this then only resulted in the addition of more definite matter of the same kind to the original conclusions, and removed all doubt. Thus the first discovery by Davy and Berzelius that electric forces must be ascribed to the atom, indeed, that atoms are held together in the molecule by electric forces – in other words, that the chemical forces are electric forces – remained all these years a good indication for research, only forgotten at times to the

[1] See Sir E. Thorpe's *History of Chemistry*, London, 1909, vol. i, p. 114.

disadvantage of the latter.   It may actually be regarded as the beginning of our present knowledge of the internal characteristics of atoms.

THOMAS YOUNG (1773–1829)
JOSEF FRAUNHOFER (1787–1826)
AUGUSTIN FRESNEL (1788–1827)

THE time had now arrived for a great advance in the knowledge of light.   Fraunhofer in Munich, and Fresnel in Paris, were responsible for this in different ways, while Thomas Young in England had to a certain extent paved the way for their discoveries.   Fraunhofer was the first to construct optical appliances of the highest refinement, and he also pointed out new methods in the art of manufacturing them, so that this art could thenceforth develop and spread on a sure basis.   His new optical apparatus belonged, however, not only to types already known, such as his astronomical telescopes of hitherto unattained power, but also to those of an entirely new description, such as his optical grating, which enabled him to attack the problem of the nature of light in an entirely new manner.

It was Fresnel who rendered the wave nature of light, imagined by Huygens and later by Thomas Young, a certainty, and he is therefore to be regarded as the founder of the wave theory of light.   He was a fine observer of recondite phenomena of light, in the same way as Newton had already observed quantitatively the colours of thin plates.   He however did not merely investigate such phenomena as had hitherto been the subjects of explanation by the wave theory with further regard to this possibility, but actually experimented with the light waves themselves, in such a way that their existence or non-existence could be subjected to decisive tests by means of a series of properly planned experiments. Just as, centuries previously, the physics of matter had gradually proceeded from observation of what offered itself more or less without preparation to thorough and well-planned experiment, so now was a similiar transition accomplished in respect of the physics of the ether founded by Huygens.

But for the fact that the domain of the ether, the basis of all known and unknown phenomena of light, electricity, and magnetism, is obviously so infinitely more extensive than that of matter (which is certainly nothing more than a peculiar and tangible formation of ether), we might call Fresnel the Galileo of the ether; but while Galileo was already able in his experiments to proceed very far towards a knowledge of the general behaviour of matter, we feel to-day as regards the ether, even after the work of Faraday, Robert Mayer and Hasenöhrl, that we are still only quite in the initial stages of knowledge. In any case, we owe to Fresnel the assurance that the ether is able to form or produce or carry waves, which are propagated with the velocity already measured by Roemer. The length of these waves—measured from peak to peak or valley to valley, without our yet being able to say what peak and valley actually mean, other than states of opposite nature[1] – was determined by Fraunhofer for the first time with very considerable accuracy by means of a new optical apparatus designed for this purpose, the grating. We are provided with a special organ, the eye, for the detection of these ether waves.

[1] Fraunhofer, in a manner very characteristic of him as a true investigator of nature, both in definite regions of known phenomena and in unknown, says himself, concerning the wave nature of light: 'Whatever we are to imagine under this term, it must in any case be of such a nature that one half of the same is opposite as regards its effect to the other half, so that if one half meets with a half of the opposite kind, the effect is eliminated, whereas it is doubled when two like halves act together in the same sense. This is the fundamental notion of interference.

The short lives enjoyed by Fraunhofer and Fresnel coincided almost to a year; accordingly their work is practically contemporaneous.   In another respect also they were alike, inasmuch as neither had any previous training whatever in preparation for their extraordinary achievements; it is thus all the clearer that these resulted entirely from their own innate gifts, urged on by which, they acquired the necessary preliminary knowledge on their own account.

Thomas Young, who was about fifteen years older than Fraunhofer and Fresnel, had studied chiefly medicine at several universities, and then became a doctor in London. This however was by no means sufficient for his active mind, and as a rich man he was always at liberty to follow up his manifold interests.   Among these was the study of the writings of Newton and Huygens, and this led him to interpret Newton's elaborate observations on the colours of thin plates and on diffraction, according to Huygens' assumption of the wave nature of light.   In this, by making use of the knowledge of sound and water waves already provided by Newton, he succeeded so excellently, that Fraunhofer and Fresnel found in the writings of Young a well prepared foundation for their further investigations.

Fraunhofer was born at Straubing on the Danube, as the tenth child of a master glassmaker, who lived in poor circumstances.   An orphan at an early age, he was apprenticed at the age of twelve to a mirror maker and glass polisher in Munich, where he also had to be of service in the house and kitchen, being bound for six years, since he could not pay a premium.   After two years the house of his master collapsed, and Fraunhofer was buried in the ruins.   As by a miracle he was extricated uninjured from the ruins, and this event so excited the sympathy of the ruling Prince, that the latter made a handsome present of money to the poor and weakly boy; he also received support in the matter of books, which his master did not possess. Fraunhofer used part of the present to buy his freedom from his master, and with the other half he bought a glass grinding machine. He also learned engraving on metal, in the hope of being now able to make himself independent. But he was not successful in earning his living, so that he could find nothing better than to return to his former employment. Five years later he was taken into a larger optical works, which he himself then helped to become famous. He soon improved all the machines, but also most particularly the manufacture of glass. His efforts led to success in making large pieces of flint glass (a lead glass with high refraction and dispersion) free from striae, and also in other respects suitable for the finest optical purposes; and he also introduced an exact control over the refractive index of the glasses used, by measuring the co-efficient of refraction (according to Snell's law) by means of prisms cut from the glasses to be tested.

These efforts led Fraunhofer to two very great successes. The first was the discovery of 'Fraunhofer's lines' in the solar spectrum, which Newton had not seen; the other was the possibility of providing astronomy with much more effective and larger refracting telescopes, particularly for the purposes of measurement. We will discuss these two achievements both separately and in their connection.

Fraunhofer's good glass enabled him to produce prisms giving much purer spectra than Newton had been able to obtain. Furthermore, he fully recognised the other requirements necessary to obtain pure spectra: use of a fine slit for admission of the light, parallelism of the rays in the prism, and then along with these, observation through a telescope. The result was immediately – as so often in the case of a clean experiment – a new discovery. He saw in the spectrum of the sun the dark lines named after him, and he proved by exact measurement of the position of these lines with all kinds of variations in the experimental conditions, that 'these lines lie in the nature of the sun's light, and do not arise by diffraction, self-deception, etc.'[1]

He always actually obtained them under all conditions in the same manner as soon as a pure spectrum of sunlight or daylight was formed. He also sought them in the light from Venus, and found them there. Fixed stars, some bright ones which he investigated, showed lines differently grouped. Fraunhofer was not yet able to realise that this opened the way to chemical analysis of the atmospheres of the sun and fixed stars, to which the combined efforts of Bunsen and Kirchhoff led forty-five years later. Nevertheless the lines of the sun's light, peculiar to it, remained a starting point for future investigations. Fraunhofer avoided in his extremely modest, but, as regards everything actually ascertainable, very penetrating manner, every discussion of mere supposition. Instead, he gives us an exact drawing of these dark lines of the sun's spectrum, a drawing which remained unsurpassed in its perfection until the time of Kirchhoff. He also immediately used the lines – which he designated by letters from A in the red to H in the violet, as we do to-day – as fixed points in the sun's spectrum, which enabled him always to select quite exactly defined kinds of light, which had not been possible hitherto, when only the colour could be named.

This was of great importance when it was necessary to measure the refractive index of a variety of glass for lenses for different colours, an operation indispensable for the construction of good achromatic (colour-free) telescope lenses. More than sixty years ago, the construction of achromatic lenses from two different kinds of glass had already been begun, but success was uncertain, if the refractive index of the two glasses for different colours did not fulfil certain conditions.

[1] This statement forms the difference between Fraunhofer's discovery and a few previous observations of such lines in the form of chance phenomena varying with the apparatus, which observations, being indefinite in character, had been forgotten.

The measurement of the index by means of his lines, and the manufacture of glasses better adapted in their properties than those hitherto usual, and free from striae, was Fraunhofer's means for the construction of large telescopes exceeding all hitherto constructed, which he was able to supply to astronomy. This method has also since remained the standard, and has continually produced further success in technical optics. Huygens' telescope, with which he discovered the rings and moons of Saturn, had, like Galileo's, only simple lenses; Newton had preferred to make use of reflecting telescopes, which were free from colour effects, and thenceforth reflecting telescopes of increasing size had been used for penetrating more and more deeply into the depths of cosmic space.

Since Fraunhofer's time, the refracting telescope again acquired increasing importance, to which his very much improved mechanical construction as regards the mounting and measuring apparatus greatly contributed. By means of a Fraunhofer telescope supplied to the observatory in Königsberg, success was for the first time gained in the discovery of the parallax of a fixed star, which had been sought in vain since the time of Copernicus, and concerning which Bradley had made such great endeavours, more than a hundred years previously. Star No. 61 in the constellation Cygni was the first to show such a displacement every half year superimposed on its own proper motion (and the aberration); it only amounted to 0.3 seconds of arc, from which the distance of the star is easily calculated as being ten light years. The second star also for which a parallax was found, Vega (distant twenty-seven light years) was measured by means of a Fraunhofer telescope; this was the Dorpat refractor, which at that time was regarded as a giant telescope,[1] being four

[1] Compare in this connection, *Josef Fraunhofer und sein optisches Institut*, by A. Seitz, Berlin, 1926, pp. 78 ff. The diameter of lenses has now increased to about four times this.

metres in length, and having an objective about a quarter of a metre in diameter. We know the parallax to-day of well over one thousand stars; the one nearest to us, Proxima Centauri, a weak companion of α Centauri, is distant 4.2 light years.

In his last years Fraunhofer was especially busy with the long-known phenomenon of diffraction. It is seen in a very simple manner when light, for example from the sun, passes through a very narrow opening into a dark room, and here meets with an obstacle, for example a hair or a screen with a small opening. We then find in the shadow of the obstacle a distribution of light which shows clearly a departure from straight line propagation: the light 'goes round the corner.' This is the diffraction already fully described by Grimaldi.[1] Newton made further variation in the experimental investigation of it, and thus provided a basis for interpreting the phenomena, which are complex in their details; but Fresnel was the first to reach a proper understanding of diffraction.

Before Fresnel had completed his work, and devised the mirror experiment which proved convincingly the wave nature of light, Fraunhofer had made a great step forward on the basis of the knowledge of Young's discussion, a step which revealed an entirely new phenomenon, namely the pure grating spectrum, which in contrast with the minuteness and feeble luminosity of diffraction phenomena hitherto known, showed a magnificent development of colour, and opened up entirely new possibilities. In these spectra, the various colour components of the light falling on the grating, for example white light, appear adjacent to one another, as in the spectrum of the prism, with the further peculiarity, however, that they are arranged in exact accordance with their wave-length, so that for example the extreme red is deflected exactly twice as much as the extreme violet, which has half its wave-length. Fraunhofer's first grating

---

[1] Grimaldi lived between 1618 and 1663 in Bologna. Newton begins the third book of his *Opticks* with a mention of Grimaldi.

consisted of fine wires, stretched in large numbers exactly parallel to one another, and at exactly equal distances apart; he then scratched gratings on glass, coated with gold leaf, by means of a dividing engine, which removed the gold in parallel lines, and finally, he engraved plain glass with a very fine diamond point. The finest grating which he made in this manner had 300 lines to the millimetre. The finer the grating, the longer the spectra which it gives. Fraunhofer again found his dark lines in grating spectra of the sun's light – a further proof of the purity of the spectra, and also of the strict relation of these lines to the light of the sun. Both the grating and the prism are used to-day for spectral investigations; the first has the advantage, on account of the peculiarity we have mentioned, of allowing measurements of wave-length to be made with ease and very great exactness, and Fraunhofer was also the first to carry out such measurements with an accuracy, for example as regards his lines, which had hitherto appeared quite unattainable.

Almost a century after Fraunhofer's time it was discovered that the newly-found and extremely short ether waves (X-rays), could be investigated by means of gratings; a grating of suitable fineness was provided by crystals, which have their atoms arranged at equal distances apart (X-ray spectroscopy). Fraunhofer scarcely allowed himself any rest in his scientific work, and particularly in the work of his optical factory, and even when he fell ill in 1825, with inflammation of the lungs, he did not take a holiday. He was not married. He lived a very modest and simple life; but the Munich Academy paid him great honour. He did not recover from the disease; he died when only thirty-nine years of age. His grave in Munich bears the epitaph *Approximavit sidera* (He brought the stars nearer). One of his biographers[1] praises in him, summarising excellently what we see from his work and his life, 'profound perspicacity,

---

[1] E. Lommel, *Gesammelte Schriften von Josef Fraunhofer*, Munich, 1888.

powerful inventive genius, tireless industry, strict love of truth, and technical mastery.'

Fresnel came from Normandy; his father was an architect.[1] He made poor progress at school; he learned slowly and with great difficulty. Nevertheless, when he was able to go to the technical high school in Paris, at the age of sixteen and a half, he began to make rapid progress, especially in the mathematical sciences. At about the age of twenty, he had completed his education in road and water engineering, and he then entered State service in this capacity. This remained his chief occupation throughout life. The year 1815 brought a short interruption, as Napoleon returned from Elba, and Fresnel was most anxious to join the troops which were to oppose his entry into Paris, in spite of the fact that Fresnel was little adapted to bodily exertion. He was taken a prisoner half dead, but was treated with great gentleness.

At this time he began his optical studies, to which he was led by reports from the Academy in Paris, and these studies occupied his thoughts when engaged on the lonely occupation of supervising roadmaking. He read at that time of 'polarised light' and complains in letters that he cannot guess what it is. When he was referred to books (such as Hauy's *Physics*) he soon was so far advanced as to be able to undertake his own investigation with the aid of simple apparatus in his leisure hours, and in the same year, 1815, he presented to the Paris Academy his first and valuable paper on the diffraction of light. From this time to the year 1826, appeared the great series of his profound optical researches, which are not only astonishing even to-day, but have actually retained their full value as regards their innumerable results. The cost of his necessary experimental equipment was met by his entering the Commission for lighthouses, and obtaining the position of an examiner in the technical school; but these were

[1] Compare *Œuvres Complètes de A. Fresnel*, Paris, 1870.   3 vols.

activities which demanded a great deal of him, and he was not able to obtain lighter work. In the year 1824 his health was undermined; he began to suffer from hæmorrhage of the lungs. His life, like that of Fraunhofer's, only lasted thirty-nine years.

Fresnel's investigations started from simple cases of diffraction, which were already known, though they are fairly complicated even in the case of the shadow cast by a single hair or aperture. These he investigated in all their details with innumerable variations, and in the finest manner possible, each experiment representing a definite question put to Nature. He found that the phenomena entirely corresponded to a wave motion according to the principle already put forward by Huygens, if we take into account that waves can always exhibit interference. Actually, all phenomena of diffraction could be explained as due to interference of the diffracted light. The diffraction itself, the possibility of propagation in curved lines, is already contained in Huygens' principle.

The further phenomena of interference were also in themselves not new, since they had already been taken into account by Sauveur as regards sound, in the time of Newton; they depend upon the fact that in every train of waves, no matter what kind, opposite states follow one another regularly, as crest and trough in the case of water waves, condensation and rarefaction in the case of sound waves, the distance apart of these states being half a wave-length. If two trains of waves meet at a point, having opposite states of equal length, they must annihilate one another, and this mutual destruction of two trains of waves is called interference. When like states meet, the trains of waves strengthen one another. In the case of sound, beats, which result from the meeting of two notes not exactly equal in pitch, had already been thoroughly explained by Sauveur as an interference phenomenon. If light is a wave motion, as Huygens already thought, it should also exhibit interference, and this arises in the case of diffraction as a result of differences in path length from the various illuminated points of the opening producing diffraction, to the screen; half a wave-length difference in the path always gives opposite states of vibration and hence darkness, or destruction of the light at the point in question of the receiving screen. The colours arise from the difference in the wave-length of the lights of different colour. If, for example, red is blotted out of white light, the remainder appears green, as had been clear since Newton's study of colour. In this way, the colour phenomena even allowed the wave-lengths of the lights of various colour to be calculated.

Something of this kind had already been attempted, before Fresnel's time, by Thomas Young, in the case of Newton's rings, which he interpreted as a phenomenon of interference, Newton's measurements only needing to be suitably interpreted in order to allow the wave-lengths to be deduced. Young had also interpreted the diminution in the size of the rings, obtained by Newton when he filled the space between his glasses with water, as a reduction of the wave-length in water. He had also concluded from the diminution in wave-length, according to the ideas already developed by Newton, that the velocity of light is smaller in water, and he had shown from Newton's figures that this diminution exactly corresponds in amount to Huygens' explanation of the refraction of waves, so that everything thus corresponds in the best possible way with the assumption of the wave nature of light. Young had also been able to explain the black spot in the middle of Newton's rings, the cause of which is not immediately obvious, by means of an acoustic analogy, and had tested this explanation by special and finely devised experiments, and confirmed it.[1]

---

[1] A good and well-informed critical discussion of the achievements of Young, and also of those of Fresnel, Grimaldi, and Malus, is to be found in E. Mach's *Physical Optics*.

Fresnel then built further on this foundation, which however led, in the case of interference phenomena, to very difficult calculations, which could never have been carried out without the calculus of Newton and Leibniz. But he could show that the explanation of diffraction by the wave hypothesis is completely successful, and that it suffices without exception even when the experimental conditions are greatly changed. In this matter, and also later, Fresnel worked in part together with Arago (Member of the Paris Academy, lived 1786–1853), whose goodwill he had won through his first communication on diffraction to the Paris Academy.

But Fresnel was not satisfied with a complete explanation by means of the wave theory of diffraction phenomena and the colours of thin plates, but wished to test this assumption on his own account in as clear an experiment as possible. This would have to be an interference experiment of the simplest description without any diffraction, in which two rays would simply be exposed to measurable differences of path, but not treated otherwise differently, so that alternate extinction and intensification as the difference in path grows greater, could be taken directly and exclusively as a sign of the presence of periodically changing opposite states along the ray, that is to say, a wave phenomenon.

Newton's experiments with thin plates appeared almost as simple as this; here however, a bundle of rays is split up into a reflected and a refracted path, which is later reflected, and the possibility cannot be dismissed, that the periodically changing states in the ray already assumed by Newton, might be states of easy reflection and easy refractio: , and not states of opposite nature and hence capable of mutual annihilation when they meet, as would be the case with a wave. Such an interpretation was excluded in the case of Fresnel's mirror experiment, which has with justification become famous, since here the two parts of a bundle of rays are reflected at two perfectly equal mirrors which meet at an obtuse angle, so that the two parts after reflection pass through one another, and therefore are able to produce interference. The experiment was completely successful in the manner expected. Interference bands were seen in the whole space where the two reflections met. This mirror experiment produced conviction on all sides regarding the wave nature of light. It also resulted in Thomas Young's explanation of Newton's rings being shown to correspond with reality.

Altogether, it was thenceforward necessary to regard a ray of light as a wave propagated with the velocity of light as measured by Roemer, a wave in a medium consisting in what Huygens had already called ether, and not of the kinds of atoms out of which, as was known since Dalton's and Berzelius' time, the ordinary matter of chemistry is made. This result has been subsequently confirmed without exception, although the ether itself is still full of questions. The lengths of the waves could also be measured in the simplest and most convincing manner by Fresnel's mirror experiments. The exactitude could not, however, be driven in this case nearly as far as with Fraunhofer's gratings, but the phenomena of these, since they are more complicated and connected with diffraction, are correspondingly less suited to provide simple and direct insight. These wave-lengths are somewhat different for different colours, but very small throughout the whole visible spectrum; they amount from the violet to the red, only to from four to eight ten-thousandths of a millimetre.

Fresnel also devoted very thorough and important investigations to polarised light and its production. Light of this kind, having unsymmetrical properties, but not directly perceptible to the eye as peculiar, had first been discovered by Huygens, as resulting from double refraction, in which 'natural, unpolarised light' was divided into two polarised rays. Not until more than a century later (1808) did Malus in Paris, who had been a soldier in Napoleon's army in Egypt, but who was fond of scientific studies, find that such polarised light may also be produced by ordinary reflection at a glass or water surface. Reflection is likewise also a means of recognition of the unsymmetry of such a polarised ray of light. If, for example, such a ray is travelling vertically, and can be reflected by a suitably inclined mirror in a forward or backward direction, the reflection fails in the other two directions, to right and left. From this it is directly deducible, that such a ray is not similarly constituted in every direction, one of the possible directions at right angles to it being in some way preferred. If we are convinced, as by Fresnel's mirror experiment (which also succeeds with polarised light), that a ray of light is a train of waves, the polarised ray cannot possibly be a train of waves vibrating in the direction of propagation, as is the case with sound; for if an alternating change of state of any kind existed in the direction of propagation only, the properties of such a ray would be similar in all directions at right angles to it. Hence the polarised ray must oscillate transversely, and further, when it is completely polarised, in only a single direction at right angles to the ray. It still remained entirely unknown what it is that changes periodically into its opposite in this transverse direction, and in the other transverse direction at right angles does not change at all.

Fresnel's recognition of the nature of polarised light became clearer, the further he penetrated into the study of its properties. It is noteworthy that not a single one of the most eminent members of the Paris Academy at that time – neither Laplace nor Arago – was able to follow him. Indeed, Fresnel himself appears to have had difficulties with the idea of transverse oscillations; but he nevertheless held firmly to the actual signs given by nature, and followed them – with complete success, as the future showed. The difficulty lay in the fact that transverse waves only occur in solid

Ps

bodies, and not in the interior of liquids, let alone gases. So how could they be possible in the ether? This difficulty was of course in reality only an imaginary one; it arose from an unclear desire to ascribe the properties of matter to the ether, although as Huygens had already expressly stated, it could not be matter.

In this case, and still even to-day, it is a fundamentally crude materialism which stands in the way of an understanding of nature; the belief then and now that everything must possess the properties which have become familiar to us from a study of matter, which study was at that time fairly far advanced. If the investigation of nature – and at the same time our whole mental development – is not to become infertile, it is time that the opposite view should again be regarded as self-evident; that the laws of matter, or mechanics, must remain entirely limited to matter, and that the ether and everything belonging to it form a part of the world of a different nature, the peculiarities of which must be first discovered piece by piece, in which process we shall meet with surprises without end, as far as investigation is able to come to grips with it. The desire to refer everything in the world to mechanism, and the feeling of satisfaction whenever this appears possible, was firmly ingrained in the Paris Academy from the time of the Encyclopaedists. It received a fresh and striking expression in Fresnel's time through Laplace's *Mécanique céleste*. This work appeared to lay down the principle that everything could be referred to the laws of motion of matter as settled by Newton, together with Newton's law of gravitation, the similar laws of Coulomb, and the molecular forces (which were also to hold for light and heat, themselves regarded as simple substances).

Fresnel had somehow or other to find a place for himself in this theoretical structure, for the Paris Academy was his only refuge. But his teacher in his experimental studies of light was Nature. In order to pacify the Academy (perhaps also himself, in a certain sense?), he stated his deduction of the transverse nature of polarised light accompanied by a hypothetical explanation as to how transverse waves might still be possible even in ordinary liquids. This mechanical explanation of transverse ether waves later became of no importance: after the results obtained by Faraday, Maxwell, and Hertz, it has become clear also in detail, that the ether is a thing by itself, not only as regards its intangibility (its incapacity for being confined in vessels).

It is not superfluous to point out these peculiar difficulties of Fresnel's time. It was the time of greatest departure of the study of nature (and altogether of humanity's world picture) from reality, as a result of complete submission to matter. Kepler, like Newton, was still free from this tendency. Huygens was not; for he says at one point[1] of 'true philosophy,' that in it 'the cause of all natural effects is found in mechanical reasons.' Perhaps that was also the cause of his not discovering gravitation, although no one at that time was nearer to it than he. Newton calmly announced his law of gravitation, although he had no mechanical explanation of it to offer. Newton also remarks concerning polarised light, a long time before Malus, and solely from the study of Iceland spar as with Huygens, quite calmly: 'Has not the ray different sides with different properties?'[2] and he remarks that this would not be possible in the case of a longitudinal or pressure wave. The conclusion as to transverse waves had to be drawn, when Fresnel had proved the existence of waves in the ray.

Fresnel then sought to discover the wave nature of ordinary unpolarised light. In view of the complete absence of unsymmetry, it might be longitudinal; according to the manner in which polarised light was derived from ordinary light another supposition was suggested: that the ordinary

[1] At the beginning of his essay on light, 1690.
[2] Newton, *Opticks*, 1717, third book, query 26.

light coming from the sun and artificial sources, likewise operates transversely, but with extremely rapid change between all possible directions of oscillations at right angles to the ray, while only a single such direction is present in the case of polarised light. This supposition has always subsequently proved satisfactory. It also allowed Fresnel to render the complicated phenomena of crystalline optics, which Arago had in part observed, comprehensible in a perfectly satisfactory manner.

Fresnel also succeeded in answering, in formulae which still arouse our admiration to-day, the question of the intensity of the reflected and transmitted light in ordinary refraction, together with the associated question of polarisation on reflection and refraction. Finally, we must mention his highly important calculation of the 'light-drag,' that is an influence to be expected of the motion of the material medium (for example, water or glass) through which light travels, upon the velocity of the latter. This calculation was also shown to be correct, when measurements were made first by Fizeau in 1853, in Paris, and later by others with much greater refinement. It may astonish us that calculations, such as that of the intensity formulae or the light-drag, should prove correct; for they treat the ether to a certain extent as an elastic material body, which it is not, but with which it has a certain similarity. But the question is simply whether the suppositions go beyond the actually existing similarity, and the success proves, that Fresnel had succeeded, here also, in doing the right, thing, without doubt as a result of his insight into the phenomena of light.

Fresnel laid the complete foundations of the wave theory of light, and also himself built upon them extensively. What was added after his time was on the one hand the treatment of special cases, for instance phenomena of diffraction, in which F. M. Schwerd (schoolmaster in Speyer, 1792–1871) did excellent work. This provided still further proof of the

HANS CHRISTIAN OERSTED

PIERRE SIMON LAPLACE

adequacy of the theory; while on the other hand, the discoveries of Oersted and Faraday afforded entirely new insight into the peculiar nature of ether waves: they were shown to be electro-magnetic in nature.

HANS CHRISTIAN OERSTED
1777–1851

OERSTED was the discoverer of the magnetic effects of electric currents, and thus of the connection between electricity and magnetism. For over a thousand years these two phenomena of nature had been known and regarded as similar, but nevertheless distinct, and without any relationship to one another. Gray had recognised the non-interference between electrical and magnetic forces acting simultaneously on the same bodies, and the conclusion that they were completely independent also remained after more refined observations had been made. Later, nevertheless, when magnetisation by means of electric sparks and lightning had been observed, and Coulomb had discovered the laws of both forces to be completely alike, the supposition of a hidden connection reappeared again and again, and in this connection was energetically sought for without success.

When Volta's battery became known as a new source of electricity, the attempt was repeated with its aid. Several experiments were published, to show the presence of magnetic force in the pile, in elements in the cup form, or also in the contact of two different metals. They were obviously in part rendered worthless by mistakes; a serious experiment for example was the following, to hang up a Voltaic pile on a silk thread, so as to move horizontally, in order to see whether it would set itself in the meridian like a magnetic needle, which of course did not happen. Efforts of this kind show that attention was paid not so much to the new possibility of obtaining a plentiful flow of electricity by Volta's invention, but rather in a confused sort of way, to the new manner of producing static electrification.

Oersted was obviously the first to investigate seriously the pile with the current flowing.   The discovery was then easily made; he saw that the magnetic needle moved when brought close to a conductor carrying a current.   It is quite incorrect to describe this as a chance discovery, for the search for a connection was being made, and not even by Oersted alone; hence the simultaneous presence of the pile and the magnetic needle on Oersted's laboratory table, and the observation of their effect upon one another was anything but chance.   The fact that the discovery took place during a lecture is, assuming it to be so,[1] not remarkable, when we remember that Oersted delighted to hold a great many lectures, even as many as four in one day.

We may also remark here quite generally, that it is a sign of widespread ignorance of the matter, when the credit due to a discoverer is regarded as diminished by his not having really known beforehand what was to be found.   On the contrary, a discovery is always the more remarkable, the further it reaches out into the completely unknown, the less therefore it is possible to see further ahead than that something was to be sought in a certain direction.   He who does not know how to judge discoveries in this light, regards Mother Nature as being as poorly endowed as our own human mind.   Every great discoverer has been surprised

[1] It is difficult to be clear from Oersted's short Latin publication (for neither Oersted nor ourselves were born to speak Latin) whether he wished to state that the discovery took place during a lecture in the winter of 1819 to 1820, or rather that the first experiment could already be seen at this time at one of his lectures, that is to say by a number of people.   The latter is more probable, when we consider the overwhelming novelty of the positive result, and the ease with which it was obtained, once one knows how, together with the widespread search going on at that time, and the delay which was inevitable before printed publication of the detailed investigation could take place.

by his own discovery.[1] What may diminish the credit due for a discovery is – apart from want of understanding of its importance, which was by no means the case with Oersted – mainly the circumstance of its having resulted from appliances previously created by others, appliances which to a certain extent already contained the discovery hidden in them, and only needed to be used in a somewhat different way, in order that the discovery might be brought to light. For this reason, we must place Galvani and Volta alongside Oersted. All the same, it took no less than twenty years before Volta's battery caused a magnetic needle to move before eyes which could understand the matter.

Immediately after the first observation, Oersted obtained, with the help of some of his learned friends, a larger 'galvanic apparatus,' consisting of twenty voltaic elements, each a large copper trough filled with dilute acid and with a zinc plate dipping into it, in order to make a further study of the effect under as favourable conditions as possible.[2] He determined in detail the changes in direction of the magnetic needle with different positions and directions of the conductor, above it, below it, sideways, and at different distances; the result was the discovery of all that we can state even to-day, without going into the matter quantitatively. He saw clearly that the newly discovered force emanating from the conductor was not an attraction or repulsion of the magnetic poles, but that it is directed in circles around the conductor. Conductors of eight different metals were tried, and found practically equal in effect, but the path of the current must not be interrupted, say by too long a layer of water. The effect is operative through glass, metal, wood,

---

[1] For example, Davy of course knew very well that he was looking for metals in the caustic alkalis, but the peculiar properties of the metals potassium and sodium which he found were no less new and surprising to him than to everyone else. See also *Nature*, Aug. 29, 1931, Article by Kirstine Meyer, 'Faraday and Oersted.'

[2] He later announced that so many elements are not necessary, a fact obvious to us to-day from Ohm's law.

water, resin, clay, and stone, when they are placed between the conductor and the magnetic needle. The needle could be placed without change in the effect in a brass box, and this could be filled with water. Needles of brass, glass, resin, and other substances, when suspended so as to be capable of motion like a magnetic needle, remained uninfluenced.

A short description in Latin of all these and other experiments was sent by Oersted at the end of July 1820 to many learned societies, single individuals, and periodicals. The discovery thus very quickly became generally known. It produced just as much general astonishment as Volta's pile twenty years previously, and though at the time the astonishment was somewhat superficial, the subsequent history up to to-day has shown that almost everything founded and achieved from then on by scientific research and technology has come to us from Volta's pile by way of Oersted's discovery, with the sole exception of the series of ideas derived from Rumford, Carnot, and Robert Mayer, which attacked the study of nature at another point, and went still more deeply.

Oersted was born on the island of Langeland, where his father was an apothecary.[1] His family circumstances were not prosperous, for which reason little could be spent on schooling. However, Hans Christian taught himself arithmetic from an old schoolbook, and also much else by an exchange of ideas with his brother, who was a little younger than himself, and by the occasional help of private teachers. At the age of twelve his father took him to assist in the shop, where he soon found great pleasure in chemical work. For this reason he was very anxious to go to the University, and entered for the matriculation examination, which he and also his brother succeeded in passing. At the age of seventeen he went to Copenhagen, where his studies – combined

---

[1] Oersted's life is described in a preface to a translation of his book *The Soul in Nature*, by L. and J. B. Horner, London, 1852.

with an extremely modest mode of living – were of the widest possible description, but principally concerned with science, philosophy, and medicine. At the age of twenty-two he obtained his doctorate in medicine. At the same time he was allowed to give lectures on chemistry and metaphysics, and took over the management of an apothecary's business.

At this time Volta's discovery became known, and Oersted immediately commenced to take an interest in it. Later he made an extensive journey to many German University towns, where he made the acquaintance of numerous eminent contemporaries, who were quickly won over by his lively mind, his youthful freshness, and his almost childlike appearance and behaviour. In the year 1806, he became professor of physics at the University of Copenhagen, where he continued, with few interruptions, to deliver a great number of lectures, also public ones. Before he made his discovery, he made a second long tour through Germany, and to Paris, and married. The discovery brought him great honour, and also many endowments and prizes, and from that time forth he was one of the most eminent and influential personalities in his own country. In course of his public activities, which later on increased, he joined the so-called 'Liberal' or 'Free Thought' party, which grew up at that time. He then undertook a third long tour to France, England, Norway, and again to Germany, where he met Gauss. We must also mention his measurements of the compressibility of water, which were executed in his later years. At his desire he was presented, as a place of retirement, with a country house surrounded by a park, but before he could move into it, he died at the age of seventy-four.

Oersted's discovery was taken up at once by a great many people. Above all, it immediately afforded a good method of measuring electric current; all that was needed was a fixed

piece of wire passing beneath a magnetic needle set in the
meridian, to enable stronger currents to be estimated quantitatively according to the amount of the deflection of the
needle. Oersted's publication already showed the direct
way to deal with weaker currents, namely to bend the conductor in a suitable manner several times round the needle,
which led, already in the same year 1820, to the 'multiplier'
provided with many turns of wire surrounding the needle
(Schweigger in Halle, and Poggendorff in Berlin). The
single turns of wire were at first insulated with resin or
sealing wax; but silk covered wires were soon introduced.
Among all those who immediately busied themselves with the
new field, Ampère was pre-eminent by the depth of the
ideas and the importance of the further discoveries, which he
developed.

PIERRE SIMON LAPLACE (*1749–1827*)
ANDRÉ MARIE AMPÈRE (*1775–1836*)

THE French Academy, which had already produced work of
such importance in the investigation of light by Fresnel,
again placed itself at first in the forefront, thanks to Ampère,
in respect of Oersted's discovery.   The high reputation
of this Academy was certainly favourable both to Fresnel's
and to Ampère's activity, and this reputation it owed to a
number of great intellects among its members, over whom
presided Laplace, at that time seventy years of age, and on
the pinnacle of fame.

Laplace came from poor parents in Normandy. He attended a military school, but was soon noticed at a very early
age in the circles of the Paris Academy on account of his
eminent mathematical gifts, whereby he attained to increasingly important offices, with influence on education, and

also when very young, to membership of the Academy.
During his life, which coincided with France's most change-
ful and fateful years, Laplace, who had ambitions in the
direction of leading activity in the state, always managed to
ingratiate himself with the coming holder of power. Thus
his influence increased, independently of the fate of his
people, right into his old age.

Laplace's achievements as a man of science are above all
concerned with the first advances towards full understanding
of molecular forces in liquids. He is the founder of the
theory of capillarity, by which name much more is designated than an understanding and quantitative knowledge of
the rise of liquids in narrow tubes, which Leonardo had
recognised as of especial interest. To this category, for
example, also belongs the formation of drops, indeed everything connected with that self-acting assumption of shape
which is a peculiarity of liquids, and which is most strikingly
shown in the case of small quantities of liquid, where the
force of gravity plays a less part, as Galileo also fully recognised. Laplace opened the road to an understanding of all
these phenomena from the quantitative point of view, by
referring them to the attractive forces of the parts (molecules) of all bodies, which forces act only over very small distances, and had already been recognised by Newton as distinct from gravitation.[1] He thus arrives for the first time at a
complete explanation of the inverse proportionality between
the capillary rise in tubes and the diameter of the latter, a
fact already known from observation on the rise of water,
Laplace's explanation depending upon the effect of the
molecular forces. Attention had hitherto been diverted
from this by Newton's idea of the very small range of
action of these forces; but it was only possible to make use of
this idea by the application of the greatest mathematical
skill, which Laplace possessed, using the infinitesimal

[1] This is to be found in the fourth volume of his *Mécanique céleste*.

calculus invented by Newton and Leibniz and since still further developed.

This calculus had already been of extensive service to scientific research. Even before the time of Scheele and Watt (from 1759 onwards), the development of the differential equations of hydrodynamics, for example, was begun. Here the answer to all questions whatever concerning the motion of liquids is referred to the mathematical treatment of these equations, which themselves, however, contain only the fundamental laws of Galileo and Newton concerning all motion, applied to the single space element of the liquid. All that is required in addition is certain equations which contain special and simple facts of experience, such as the small compressibility of liquids and the like, and also special statements relating to the case considered. Though these equations, when skilfully handled, answer questions which might otherwise appear insoluble, for example those relating to all details of waves upon liquids, and though they go in their generality far beyond the consideration of characteristic and hence important cases already dealt with by Newton, they nevertheless contain only old knowledge; nothing fundamentally new is added or made evident by their treatment. Nothing else can possibly be the case; for mathematics is throughout scientific research, simply a tool, enabling us to apply knowledge derived from the observation of accessible and generally simple cases, in a completely logical manner and without danger of mental error, to all other cases whatsoever, no matter how complicated; but knowledge of nature can only be derived from observation.

It is the special virtue of the art of mathematics that, when correctly applied without arbitrariness, it keeps away all foreign matter, and only allows that to take effect which was originally derived from observation and put into the equations. If the final result then leads to matters which excite astonishment, as when for example Laplace concludes that

Andre Marie Ampère

SADI CARNOT

the solar system, in spite of all disturbances by the mutual gravitation of the single planets and moon, will never at any future time become disordered, these are nevertheless no fundamental advances in knowledge. For they are only exactly as true and correct, that is to say in agreement with reality, as what was originally put into the equations, for example in the case mentioned, Newton's law of gravitation and the laws of motion of matter, discovered by Galileo, Huygens and Newton, and only in so far as nothing unknown and therefore not taken account of in the original equation, plays a part.[1] In such cases we may say that mathematics is able to give us results that we did not know that we knew, that is to say, that we already possessed the necessary knowledge of nature required to determine them.

This peculiarity of mathematical achievement in the questions of science is but little accessible to the general mind; it is therefore often overlooked, and mathematicians are thus often confused with scientific investigators. Mathematical skill can, of course, bring about advance in scientific knowledge; but this only happens when the discovery and observation of natural processes which are still unknown, or of a new kind, or not yet properly understood – the highest achievement of the experimental investigator – yet fails to bring full understanding of the processes by means of simple considerations, by reason of their complexity.

Thus Newton became by means of mathematical skill the discoverer and founder of the law of gravitation, by

[1] Thus Laplace's much admired result concerning the absolute stability of the solar system (which Newton still regarded as quite questionable and even improbable) would be completely upset, as soon as any sufficiently large cosmic body approached too close to the system from outside, and there are far more dark masses having unknown motions in cosmic space, than was earlier assumed by astronomy. Should, however, the results of the stability of the solar system be thus upset, this would not lie in the want of validity of Newton's laws of gravitation and Galileo's laws of motion, which are assumed in Laplace's calculation; the calculations would be inapplicable, but not the facts on which they were founded.

examining the complicated motions of the planets, and calculating these down to their details. But the description we have given, which was entirely limited to essential points, must have shown the reader in this case also that the discovery itself – though not its confirmation in all directions – nevertheless resulted from comparatively simple lines of thought, which also needed only fundamentally simple mathematical means for their support. The same will have been noticed in all revelations of new and unsuspected secrets of nature, such as are characteristic of the whole series of great men of science hitherto considered: indeed, the discoveries which most of all exceeded all previous possibilities, and most strikingly overthrew existing limits of knowledge, such as those of Volta, Davy, and Oersted, did not need the help of mathematics, and the same is true up to to-day. The importance and effectiveness of mathematics in scientific research is quite generally overestimated. Investigation, it is true, must always strive to be quantitative, but the fundamental relationships of a quantitative nature which hold in fact, and the discovery of which is the sole object, have always proved to be of the simplest possible description.

Laplace's great work, the *Mécanique céleste* (celestial mechanics) was for the most part a purely mathematical achievement. The work starts from the discoveries already described of Galileo, Huygens, and Newton, and develops extremely valuable mathematical methods for applying them in a far more detailed and accurate manner than Newton, to the motion of the planets, the moon, and the waters of the sea. Laplace thereby developed valuable mathematical methods for calculating the disturbances of planets and moons by their mutual gravitational forces, as well as quite general methods for calculating forces which act according to the inverse square law (the theory of potential).

A second scientific achievement of Laplace, is his discovery of the fact that, and of the degree to which, the

velocity of sound is influenced by the heat phenomena which
are necessarily always an accompaniment of the sound
waves.   Dalton had already showed that air becomes heated
when compressed, and cooled when expanded, and Gay-
Lussac made the first measurements of this phenomenon.
But pressure changes of this kind take place in all sound
waves, and the temperature changes connected therewith
cannot be without influence on the velocity of propagation,
for they influence the elastic forces of the air.   Laplace
calculated this influence, using as his basis Newton's first calculation of sound velocity, and found that we need to add to
Newton's formula the ratio of the two specific heats of air
(at constant pressure and constant volume) under the square
root sign.   The more accurate determinations of the velocity
of sound then available already demanded such an increase as
compared with Newton's formula, and the refined measurements of the ratio of the two specific heats which were then
undertaken – on the basis of a sharp definition first given
by Laplace in this connection – showed full agreement
between Laplace's calculation and hence the idea upon
which it was based, and reality.

Ampère was born in Lyons, where his father was a well-
to-do merchant.   He early showed extraordinary mathematical ability, but in other respects also a wide range of
mental activity.   He was brought up alone in the country,
to which his father had retired, and his studies were accomplished, with a little assistance from his father, by means of
books.   At this time of many sided education and growing
personality, at the age of eighteen years, he had the misfortune to be cruelly robbed of his father, who was declared
to be an aristocrat, and became a victim of the mass-murder
of the Revolution.   For over a year he wandered about
distracted and planless, until he was seized for the moment
by an especial interest in botany, one of the many studies

which he had pursued.   At that time also, he wrote poetry.
His marriage at the age of twenty-four then gave his life a
definite direction again; he settled in Lyons, where he taught
mathematics.   He later became professor of physics and
chemistry at a school in Bourg (north of Lyons); but his
family, on account of the serious illness of his wife, was not
able to follow him thither; he soon lost her, after only four
years of marriage.   A mathematical work of his then led to
an appointment in Paris, where he gradually rose to the
highest scientific posts.

Oersted's discovery thereafter decided the direction in
which his gift for scientific research was to develop.   In a
short time he unfolded the astonishingly fertile activity
which resulted in the creation of the main part of electro-
dynamics (the science of moving electricity).   While
Galvani and Volta supplied the means for this development, and Oersted showed a main road to its application,
Ampère gave this application – which led to electromagnetism – a fixed form which is serviceable to-day, and
forms the main part of its content; only Faraday was later
able to add to it something new and important, both in form
and matter.   Ampère died, sixteen years after Oersted's
discovery, at the age of sixty-one.   He, like Fresnel, was
obviously of a different nature from the majority of his colleagues in the Academy; this is shown by various traits of
character, which his contemporary biographers regarded as
peculiar,[1] and also by the fact that he, probably alone in the
Academy, did not trust the idea of caloric, and also in his
wide sympathy for all fundamental questions of natural
science, even when they related to ideas, at that time regarded as very far fetched, concerning the evolution of living
creatures.

[1] Arago, *Oeuvres, Tome II.   Notices biographiques,* 1854.   A somewhat earlier and good biography is found in the *Revue des deux mondes,*
vol. ix, page 389, 1837.

He retained a childlike disposition up to his old age, but apparently he was not without a great deal of self-will. He was often tortured by doubt in small matters as well as great ones, and at times, after the death of his first wife, suffered domestic misfortune, so that life, in spite of all the recognition that it brought him, was by no means always satisfactory, a fact expressed in the epitaph chosen by himself: *tandem felix* (Happy at last).[1]

Ampère's successes in electro-dynamics depend before all upon his clear grasp of the essential features of these phenomena, first opened up by Oersted, and presenting in the beginning a highly complicated and confusing picture, and upon his own observation. His further progress was due to skilful experiment based on questions stated with insight.[2]

Simply to grasp what happens in the connecting wire of a voltaic pile was then a difficult matter, and required a firm hold to be kept on the essential points of observation. Ampère succeeded excellently in this, and settled the idea of the electric current, at that time undefined, in a manner which has been of the greatest advantage to progress. Volta, it is true, had made use of the term electric current,[3] and people were also accustomed since Gray's time to speak of the conduction and flow of electricity. However, when we recollect that on the discharge of the pile, and likewise of a Leyden jar, both electricities were supposed to flow in opposite directions on account of their attraction for one another,

---

[1] See a collection of letters and recollections, which however does not afford very much information, *André Marie Ampère et J. J. Ampère*, by an anonymous authoress, Paris, 1875. One thing at least is clear that Ampère's doubts, which often tortured him, related to a considerable extent to the dogmas of the Catholic church, of which he was a member, as was his first wife.

[2] Ampère's publications in the period directly after Oersted's discovery were reprinted in Paris in 1921 under the title *Mémoires sur l'Électromagnétisme et l'Électrodynamique*.

[3] In Volta's original French manuscript, published in the *Philosophical Transactions*, vol. 83, 1793, we already have in the first corresponding connection the words 'courant électrique.'

Qs

whereby their well-known opposite effects were neutralised by their union, the electric current must have appeared as something fairly complicated and indefinite. The idea of it in any case became indefinite, from the fact that the simultaneous 'current' of two electricities, obviously had no defined direction. In accordance with this, Oersted did not speak of an electric current at all, but used the expression 'electric conflict'[1] for the process in the wire, and this indefinite terminology also corresponds to the confusion shown in the then rapidly accumulating publications relating to continuations of Oersted's experiment.

As opposed to this, Ampère decides in the same year, 1820, that he will call the whole process in the discharge wire an electric current, whereby no regard will be paid to details (which, by the way, are not too well understood even to-day), and that the direction of the current is to be defined as the direction in which we imagine the positive electricity to move. This made the electric current something definite: a special natural process, the observable peculiarities of which are offered as a subject for investigation undisturbed by difficulties in thought, which could not for the moment be overcome. A new and well-defined concept had been formed, and clear concepts, correctly adapted to nature, have ever been the guiding stars of progress in knowledge. In Volta's pile or cup apparatus, this current circulates in a definite direction along the closed conducting circuit, as, by the way, Volta himself also expressly states.

Ampère also distinguished sharply for the first time between the phenomena of electric tension and those of electric current. Phenomena of tension had long been known from static electricity; they appeared in the pile before the conducting circuit was closed, and they could be followed by

---

[1] The title of Oersted's Latin publication of his discovery in the year 1820 read *Experimenta circa effectum conflictus electrici in acum magneticam.*

means of the electroscope or electrometer, which instruments Ampère now expressly states to be measurers of tension. As phenomena of the electric current, on the other hand, he expressly puts forward the chemical and magnetic effects of the current (the corresponding development of heat was first really grasped by Joule), and he proves with regard to them especially, that they are not phenomena of tension, since they are completely absent when the latter alone is present (before the circuit is closed). The current, he says, is best measured by means of its magnetic effects (which is still true to-day, and for which purpose the multiplier had at that time been invented in Germany) and he introduced for every current measurer working by means of the magnetic effect, the name 'galvanometer,' which has ever since remained in use.[1]

Tension, or as we call it to-day, electromotive force (voltage), thus appears as a cause, and current as an effect. As soon as this effect results from closure of the circuit, the phenomena of tension disappear, 'or become imperceptible.' Ampère even imagined the resistance of the circuit already as determining the current strength at a given tension. We see that as regards Ampère, Ohm's law was not far off. But in the case of all his contemporaries this would be very far from true, however much they concerned themselves, together with Ampère, with Oersted's discovery; this is seen in the want of comprehension which they exhibited even seven and more years later, when confronted with Ohm's achievement.

Ampère's great step forward, taken in the year of Oersted's

[1] We may ask why not 'voltameter'? For it was not Galvani but Volta who was the first to render possible and produce the subject of measurement, namely steady electric currents, although certainly as a continuation of Galvani's investigation. I can find no other reason than perhaps the fact that Volta was still alive then, and Ampère wished to honour the dead. As a matter of fact the word 'voltameter' was introduced later by Faraday, for apparatus measuring current by its chemical effects, after Volta's death.

publication, and relating only to his grasp of the matter, immediately led him to the 'lefthand rule' still in use to-day, for indicating the direction in which a north pole is deflected by a given current. This was an improvement on Oersted's form of statement, which of course led to exactly the same result. Ampère on this occasion also introduced for the first time a clear distinction between electrostatics (the science of stationary electricity) and electrodynamics (science of moving electricity, that is of current effects), and invented these terms themselves.

To all this Ampère immediately also added an entirely new discovery. It was known that electric charges exert forces upon one another, and likewise magnetic poles – both according to Coulomb's law – and it was now known from Oersted's work, that electric currents act on magnetic poles as if they were themselves magnets. This at once gave rise for Ampère to the question whether currents might not also exert forces on other currents. He decided the question in the year of Oersted's discovery (1820), by experiments, and in the affirmative. He found that parallel currents in the same direction attract one another, and those in opposite directions repel one another, it being a matter of indifference whether the two currents acting upon one another belong to the same circuit or to separate circuits. The result was the discovery of again a new and unknown kind of force.

The fundamental experiments were of a simple description, although at that time by no means quite obvious; it was merely a matter of making current-carrying conductors easily mobile, which Ampère effected by suspending them from mercury cups and by similar means. A very remarkable fact in this connection was that here like attracted like, and opposites repelled one another, a reverse case to that of static electricity. The newly discovered electrodynamic force was thereby immediately distinguished from

the electrostatic forces, which the electricities in the wires
would exert upon one another when at rest. Ampère carried out these experiments also with wires bent into a circular
form, and with coils consisting of many circular windings,
through which he passed current. In this way coils (called
by Ampère 'solenoids') came into use, though these had
already, shortly before, appeared in the 'multiplier.' Ampère also showed that Newton's law of reaction also holds for
the newly discovered force as well as for the forces discovered
by Oersted, so that the current-carrying conductor is able to
set the magnet in motion, and conversely the magnet the
conductor.

The discovery of the forces between current and current
immediately led Ampère further to a peculiar conception of
magnetism. He regarded a magnetised rod as a coil carrying a current, a current which flows by itself uninterruptedly.
He was actually able to show that two coils, one of which is
movable, deflect one another, and act upon one another
with their ends, in exactly the same way as do two magnetised rods with their poles. Indeed he was even able to
show that a single circular current, when suspended so as to
be free to move, sets itself like a magnetic needle with reference to the earth's magnetism. In this way all magnetic
forces could be referred to the forces between current and
current; and magnetism ceased to be a thing by itself. This
was entirely in accordance with reality, inasmuch as it was
already clear since Gilbert's experiments with broken magnets, that in actual fact there is nothing peculiar about magnetic poles, but that we have in a magnetised rod a state
uniformly distributed and directed.

According to Ampère, every magnet could be imagined as
composed of circular currents similar in direction and set
side by side, and could actually be completely substituted
in all effects of its forces by such a circular current. In the
last resort it was only necessary to assume that every iron or

steel molecule of the magnet contains in itself a small permanent circular current; magnetisation would then only consist in directing all these circular currents of the innumerable iron or steel molecules in the same sense. As against this idea, and in the immediately subsequent time in an increasing degree, the objection existed, that the continued maintenance of such currents without a source of power would not be possible. This objection has only quite recently fallen to the ground, when a continued circulation of electricities in the atoms and molecules was proved to exist in quite another manner, by special facts (at first by means of Hittorf's cathode rays).

But even without these recent discoveries, Ampère's ideas bore immediate fruit. Arago had noticed that the connecting wire of Volta's pile coats itself with iron filings, which were previously entirely non-magnetic, but afterwards show permanent magnetism. He showed this to Ampère, who immediately said that a current carrying coil would turn steel needles placed in its axis into magnets, having north and south poles arranged in a manner which could be predicted. This gave the means of making much stronger magnets than had ever been possible before by the use of the lodestone. Furthermore, the way to the electromagnet was open, which retained or lost its power at will by closing and opening the current; it was only necessary to use soft iron instead of steel in the core. We know what an enormous number of applications have been found for this; the electromagnet is the main constructional element of a large range of electrotechnics, from the electric bell to the electric locomotive.

SADI CARNOT
1796–1832

THIS man of science was the first since Watt's time to bring
an essentially new idea into the science of the steam engine,
and this was an idea which he himself already showed to be
of much wider importance, namely, as holding for all apparatus which produces work by means of heat.[1] His discovery
proved of continually increasing and more general importance, until finally, twenty-five years later, Clausius made it
into the 'second law of thermo-dynamics,' which allows us to
calculate so many phenomena of heat in all their details in
the most astonishing manner. In this connection, Carnot
also arrived in his short lifetime very near to the nature of
heat itself, almost to the point of Robert Mayer's later discoveries, as we now know from his posthumous papers,
which only came to light long after.

His starting point was the question of the greatest possible
efficiency of heat engines, that is to say, the conditions
which finally determine how many horse-power hours of work
such machines, and in particular a Watt's steam engine, can
deliver, in the best case, by the consumption of a given
amount of heat in the boiler. He first realised that, in all
such apparatus, heat necessarily passes from a hot body to
a colder body, in the case of the steam engine from the
boiler to the condenser. Only where such passage of heat is
possible, that is to say, only when temperature differences
are present, can heat be turned into work. If as much work
as possible is to be done, only such transference of heat must
take place as is associated with changes in volume; for the
work is done by these changes in volume, as in the cylinder
of the steam engine, where the steam expands. Every

---

[1] Nicolas Léonard (called Sadi) Carnot, *Réflexions sur la puissance
motrice du feu et sur les machines propres à dévelloper cette puissance*, Paris,
1824; facsimile, Paris, 1912. Eng. trans. by R. H. Thurston, London,
1890.

transference of heat that takes place without change of volume, for example by simple conduction, means a loss in work.

Carnot then further remarked, that the changes in volume associated with the performance of work are reversible, whereby his ideas rest upon facts already determined by Dalton and Gay-Lussac, and likewise upon Laplace's calculation of the velocity of sound, which was found to agree with experiment. He says: 'If a gas is compressed quickly, its temperature is raised: conversely, it falls, when the gas is quickly expanded. This is one of the best ascertained facts of experience; we shall take it as a foundation of our proofs.'[1] He thus arrives at the conclusion that complete reversibility of the process used must be regarded as a condition of maximum production of work by heat. The process in a steam engine is actually reversible. For if the flywheel of the engine were turned backwards, it would act as a pump, removing the water vapour from the condenser into the boiler, whereby the first would be cooled,[2] the latter heated, so that along with the flywheel and the steam, the heat also would move in a reverse direction, from the condenser to the boiler, from the colder to the hotter body.

The efficiency of the engine depends upon the perfection of this reversibility; everything that is not reversible, such as loss of heat by conduction, frictional processes, and also the passage of steam from the cylinder into the condenser, before it – after being cut off from the boiler – has assumed of itself the condenser temperature;[3] all such processes diminish the efficiency of the engine. In the case of a

[1] In this connection, Carnot was also the first to suspect the true explanation of the temperature distribution in the earth's atmosphere, inasmuch as he remarks: 'Must not the cooling of the air by expansion account for the cold in the upper regions of the atmosphere? The reasons hitherto given as explanations of this cold are completely insufficient.'

[2] The steam engine running in a reverse direction is actually, in principle, a refrigerator.

[3] Watt already introduced the practice of cutting the steam off from the cylinder long before the end of the piston stroke, in order to make his engines operate more economically.

perfectly reversible steam engine, the efficiency could only be increased by an increase in the temperature difference between boiler and condenser.

There now only remained open the important question whether machines using another vapour than water, or generally, another body capable of doing work by expansion, such as some gas or other, might not be more advantageous, inasmuch as with the same range of temperature, and complete reversibility, differences in the power of doing work might exist according to the substance used or according to peculiarities in the construction of the machine. Carnot is able to finally answer this question in the negative by means of an imaginary experiment. He bases this – as was done more than two hundred years previously by Stevin – on our experience of the impossibility of perpetual motion. If one of the two engines compared would work more efficiently than the other, we could allow one to use part of the work produced to drive the other backwards, whereby the temperature range, used in the driving machine, would be maintained without addition of heat, on account of the action of the other machine; while in addition, the excess of work delivered by the better machine would be at our disposal. This would amount to perpetual motion. Since such an arrangement is known to be impossible, the efficiencies of the two different machines cannot differ from one another.[1]

Carnot thus concludes that all completely reversible machines acting over the same range of temperature must be equally efficient, and that their delivery of work can only depend, apart from the available quantity of heat, upon the utilisable range of temperature, increasing with both of these quantities. This discovery has been of fundamental

[1] In this connection Carnot also draws attention to the experience gained with Volta's batteries, and points out that these arrangements also can in no way be regarded as inexhaustible sources of, say, magnetic force, but always show clear signs of exhaustion, and hence cannot produce perpetual motion.

importance for the steam engine and all other heat engines since constructed; also, already without the further knowledge contributed by Robert Mayer and then Clausius, Carnot is able, at the conclusion of his essay, to add a pertinent discussion concerning possible improvements in the steam engine as a result of his work. We may also mention that, in addition, he developed very finely thought-out conclusions as regards the heat properties of gases.

Carnot's ideas are well linked together and with experience; nevertheless this connection is not completely rounded off. Carnot feels this himself, and he notes at several points certain open questions. These particularly relate to the assumption of the invariability in quantity of heat, which Carnot takes for granted, but not without remarking[1] that this assumption, which is the 'foundation of the whole theory of heat' nevertheless still 'needs attentive investigation,' since 'several facts of experience seem to be almost inexplicable according to the theory in its present state.' According to this assumption, made at that time for want of better knowledge, the quantity of heat leaving the boiler with the steam would in the steam engine be given up in undiminished quantity to the condenser, though reduced to a lower temperature, just as the water in a turbine does not diminish in amount, but does its work merely by falling in level. This is not correct, as Robert Mayer showed in 1842.[2] The heat does become less; it is partly transformed into work. It later appeared from Carnot's posthumous papers[3] that he himself had already arrived at a recognition

[1] Footnote to Carnot's essay.

[2] Two years before Carnot's publication, the invariability of quantities of heat had been successfully maintained in the large work of Fourier, *Théorie de la Chaleur*, which appeared in 1822. This work, however, which is distinguished by wonderful clarity and great mathematical power, only deals with processes of heat conduction, for which the invariability is also sufficiently true.

[3] These were first published in the year 1878 by his brother, as appendix to a second edition of the essay *Réflexions sur la puissance motrice du feu*.

of this fact, but was prevented by his all too early death from following up the matter, so that it remained completely hidden.

Sadi Carnot, who was born in Paris, was the second son of Nicolas Marguérite Carnot, known as Napoleon's minister and organiser of France's national defence. He studied technical science and entered military service.[1]

His last studies, which remained incompleted, as well as other notes by his hand,[2] show, no less than his published essay, that a rare mind was lost by his early death. We will only remark that in his papers a numerical value for the mechanical equivalent of heat (1000 metre-kilograms $\div$ 2.7 calories = 370 metre-kilograms per calorie) is given without any further indications of its origin, together with plans for a considerable part of the measurements later carried out by Joule.[3] Along with these we find remarks – as concerning the injuriousness of war to the race, religion, and everyday and purified Christianity – which expose with remarkable clearness evils still existing to-day.[4]

[1] Our portrait shows him at the age of seventeen in the uniform of a student of the École Polytechnique in Paris. Some years later, after Napoleon's abdication, which resulted in his father's banishment, he retired from the service in order to be able to devote himself, in the greatest seclusion, entirely to his scientific studies. This mode of life was only interrupted by a journey to Germany, where he visited his father who was living in Magdeburg, and later, after publication of his *Réflexions,* by a short re-entry into the army, during which he was made a Colonel. He died early, only thirty-six years of age, the victim of an epidemic of cholera.

[2] In the edition by his brother already referred to.

[3] Since Carnot's ideas had remained completely unknown, this does not affect Robert Mayer's credit for having first made this equivalent known, and generally, for having discovered the idea of it, nor Joule's credit for having carried out the experiments; but it bears of course upon the estimate we are to make of Carnot himself.

[4] These remarks may be appreciated later, when the time comes for such thinkers to be valued, even when they do not bring us only technical progress; see pages 83–87 of the 1878 edition.

GEORG SIMON OHM
1789–1854

OHM's law is to-day a commonplace for every technical electrician. It is the means by which the strength of an electric current can be calculated beforehand, when the power of the source of current and the nature of the circuit are known. When Volta had provided a plentiful source of electric current, and Oersted the means for easily measuring it, and Ampère had introduced a useful and definite conception of current, one would have supposed that a need for a knowledge of the way in which the current depends upon the factors determining it would have been felt by the many persons who now began to busy themselves with it. This was far from what actually happened. Ohm in his time announced his law to deaf ears, which shows us on the one hand how superficial the requirements of the majority were, and on the other hand how far in advance of his time Ohm was as regards clear thinking. As a matter of fact, only Davy and Ampère had approached near to the law before his time. Both of them had already grasped, in the year 1821, the idea of the resistance of a conductor, inasmuch as they took note of the dependence of the current strength upon the nature of the circuit, whereby Davy estimated the strength of the current according to its chemical effects, and thus compared a number of different metals as conductors with one another, determined the unessential effect of the form of the cross-section by comparison of simple, multiple and flattened wires, and also discovered the increase of resistance of metals with temperature.

It was Ohm who grasped the questions already illuminated by these investigators, and pursued them in all directions to complete solution. Using very poor and deficient apparatus, he performed a series of decisive experiments, which completely settled the question of the distribution of electromotive force in the circuit and also that of current strength, and he was also able to state his result in a definite manner, in the form of equations, the most important by far being Ohm's law: 'The current is equal to the driving tension or electromotive force divided by the resistance.' If this law is applied not only to the whole circuit, but also to any part of it, the distribution of electromotive force can be found in any part of the circuit.

A part of his experiments were performed with voltaic elements; in particular, he thus demonstrated the fall in potential along a long and thin connecting wire, for which purpose he, like Volta, made use of the condenser. In this way it could be proved – a question left open by Ampère, – that the voltage of the force of current does not disappear when the circuit is closed, but that it distributes itself in a definite way, in accordance with the resistances. Ohm also recognised that the sources of current discovered by Volta possess the property of maintaining a definite difference of tension at the points of contact of the different conductors, whether the circuit is open or closed, this difference in tension (electric potential) depending only on the nature of the conductors concerned. This difference in tension is also called the electromotive force of the current source in question. The fact that this is not very constant in an ordinary voltaic element, but fluctuates as Ohm says, depends, as Ohm already saw, upon the chemical changes which take place of themselves in the element when the current is used.

The constant galvanic element, which avoids these disturbing chemical changes, had not then been discovered, and this was at first a serious difficulty for Ohm. But he found a means of getting over this, by following Poggendorf's advice to use the thermo-electric elements just discovered as a source of current. This was a discovery made in the year 1821 by Seebeck (lived between 1770 and 1831 in Germany), that a circuit formed of different metals deflects the magnetic needle as soon as temperature differences exist in it. Seebeck called this 'magnetic polarisation of metals by temperature differences' or 'thermo-magnetism.' Ohm immediately regarded the discovery as the finding of a new source of current, which entirely consisted of conductors of the first class. We may also say that what had been found was really that Volta's series changes its order when the temperature is different. When Ohm introduced a piece of bismuth into a circuit otherwise consisting of copper, and kept one of the two points of contact in boiling water and the other in ice, he obtained very constant currents, which he was able to observe by means of a very simple galvanometer included in the circuit.

This provided him with a trustworthy and invariable source of electromotive force, and it thus became possible to study the influence upon the current strength of the resistances introduced into the circuit, in a trustworthy and unobjectionable manner. It was thus rendered certain that the resistance of a conductor in Ohm's law is proportional to its length, and inversely proportional to its cross-section, and to its conductivity. The connection with the cross-section shows that electricity in constant flow, as contrasted with that at rest, is not situated on the surface of a conductor, but distributed throughout its whole interior. Ohm already begins to consider the state of transition from rest to flow of electricity, when the circuit is closed, although the knowledge at that time did not allow him to deal with it satisfactorily. But he already found quite correctly that the conductivity of liquids, as opposed to that of metals, increases with increase of temperature.

Ohm came from an old Erlangen family. He received from his father who was a master-mechanic, a careful education; his mother died when he was young. During his time at school he was also introduced successfully by his father to mathematics and physics, who for this purpose, as also on account of his own desire for knowledge, studied these sciences himself from books in his old days, along with his own work and at the cost of much self-sacrifice. At sixteen Ohm then began to study mathematics, physics and philosophy at the University of his birthplace, but this does not appear to have had a very strong attraction for him. Also, money failed, so that he left the university after two years and took a teaching position in a private Swiss school. He later, however, obtained his degree in Erlangen and became a member of the university. But he was soon again compelled by want of means to leave the university. He endeavoured to obtain employment in school-teaching, and so became from 1813 to 1827 a teacher of physics and mathematics first at the Realschule in Bamberg, and then at the Gymnasium in Cologne. During the ten years at Cologne he produced the important researches we have already discussed. Since he had little free time and very poor apparatus at his disposal, the publication took place in various small papers, which supplemented one another, until a longer period of leave enabled him to produce a more connected account of them.[1]

They then appeared, without, however, any account of the experimental basis, in 1827 in an independent publication, *The Galvanic Circuit investigated Mathematically.*

Ohm was himself well aware of the importance and success of his work, and he expected some recognition, in particular desiring a better opportunity for work than could be found in schools, where his duties gave him but little satisfaction. He therefore resigned his position in the school, but had to wait for five years in very straitened circumstances until, after many petitions addressed to the King of Bavaria, he received

[1] See *The Galvanic Circuit investigated Mathematically*, translated by W. Francis, with a preface by T. D. Lockwood, New York, 1891.

Concerning Ohm's later work on optical and acoustical matters, see his collected papers (*Gesammelte Abhandlungen*), edited by E. Lommel, Leipzig, 1892.

a suitable teaching position, although not in Munich, where at the time a possibility existed, but at the Polytechnic in Nuremberg.[1] Ohm filled this post for sixteen years, during which recognition of his achievement, particularly from abroad, gradually came to him. In his sixtieth year he finally realised his desire to obtain a post in a University; he was called to Munich. Five years later he died after a short illness. He never married, and was always very simple in his tastes.

KARL FRIEDRICH GAUSS
1777–1855

GAUSS was called the prince of mathematicians. Here we do not propose to give an estimate of his mathematical achievement; we can only show to what extent he was an investigator of nature. He was one of the great mathematicians who did not deny in his work the original purpose of mathematics, that of forwarding our knowledge of nature, although pure mathematics – mathematics as a mental science – had decidedly more attraction for him.

Gauss was in actual fact also an investigator of nature throughout his life. At the age of twenty-four he already calculated the orbit of the first small planet, Ceres, discovered in 1801, which could only be observed for a short time, since it soon approached too near the sun; without Gauss' calculation it would have again been lost, that is to say not at that time have been recognised as a special planet. This successful calculation from a very short observation of the orbit aroused even Laplace's admiration; but Gauss

---

[1] Compare the detailed statements in the collection of his posthumous manuscripts made by L. Hartmann, *Aus G. S. Ohm's handschriftlichem Nachlass*, Munich, 1927.

Georg Simon Ohm

KARL FRIEDRICH GAUSS

already possessed at that time his 'method of least squares,' discovered in his eighteenth year, which allowed the influence of the inevitable errors of observation to be made as small as possible in working out results. This was the beginning of a series of discoveries of the planetoids which lie between Mars and Jupiter; up to to-day about a thousand of them have been found. The method of calculation, according to which all of them are followed in their courses, was first published by Gauss in the year 1809, in his *Theoria motus corporum coelestium* (Theory of the Motion of the Heavenly Bodies), in which he actually goes beyond Laplace's *Mécanique céleste*, which at that time had already partly appeared.

As regards our investigation and knowledge of nature, however, all this amounted to nothing more than skill in calculation according to Newton's inverse square law; none the less, in absence of this skill it would never have been so certain that this law holds without exception, and to the utmost limit of observation, throughout the whole solar system. It is just by this experimental verification of new conclusions, capable of exact tests, that all natural laws are continually put to the best possible proof concerning their agreement with reality. In this connection we should also mention the calculation – in 1846, and so within Gauss' lifetime – of the orbit of the most distant planet Neptune from the slight observed disturbances of Uranus, a particular shining example of the justified triumph of knowledge, gained on the basis of the methods of Laplace and Gauss, a hundred and sixty years after Newton had shown the way with his *Principia*.

In another direction Gauss is to be regarded as completing the work of Laplace, again on the basis of Newton's ideas, in his *Foundation of the Theory of the Form of Liquid in a State of Equilibrium* (1830) where he endeavours to attain the greatest strictness in mathematical calculation, without

Rs

which mathematics can actually become no less delusive than ordinary, and even quite superficial, thinking.

Gauss likewise developed in the same way the theory of potential already mentioned under Laplace's achievements and relating to the inverse square law; he developed this considerably further. 'Pauca sed matura' (little but ripe) was his motto, and all his publications do it the fullest honour, although their number, when we count in the purely mathematical work, is not less overwhelming than their contents. Gauss' most profound work as an investigator of nature is found in his researches on the intensity of the earth's magnetic force, reduced to absolute measure (1832). In this he founded the system of absolute units which has become ever more and more indispensable to scientific research, and which was afterwards carried out for all electrical and magnetic quantities, among which also we must reckon the technical units everywhere used to-day, – the Coulomb, the Volt, the Ohm, etc. In the experimental work he collaborated in part with his Göttingen colleague Wilhelm Weber, who was twenty-seven years younger, and whose chief life work became the development of the system for electrical quantities, together with all the preliminary work necessary for this purpose. Gauss himself confined his work to laying the foundations on the magnetic side, depending entirely on Coulomb's law for the magnetic pole, which law was also subjected to a very refined test, which it passed successfully.[1]

The system of absolute units takes its origin from three fundamental units: space, mass, and time (centimetre, gram, second); all other units are deduced from these, solely by means of laws of nature, and therefore without any other arbitrary act than the choice of these laws. The ordinary unit of force (the gram-weight) is avoided, since it only has a definite value for a definite latitude and height above sea-level,

[1] See *Nature*, Aug. 29, 1931, 'Gauss' Investigations on Electrodynamics,' by Clemens Schaefer.

and indeed away from the earth may have any value whatsoever. In the absolute system, the unit of force is derived by means of Newton's second law of motion from the units of mass and acceleration (which latter follows from the units of length and time). This unit of force is very small; it is about equal to the weight of a milligram, and later received the name 'dyne.' By means of the unit of force and Coulomb's law it is easy to deduce a unit of magnetic pole-strength. But it would be equally easy to deduce from the same law as applied to electricity a unit of electric quantity. It is here that the two branches of the absolute system of units divide: the electro-magnetic units correspond to the first choice, that of Gauss, the electro-static, to the second, which was also later followed up by Weber.

The fact that Gauss preferred to take the magnetic pole rather than the quantity of electricity, was mainly due to his desire to reduce the intensity of the earth's magnetism, as regards its geographical distribution, to a single and easily maintained unit of measurement, after comparative measurements had already been made by means of the oscillations of magnetic needles, particularly by Humboldt; and in a balloon, for the first time, by Gay-Lussac.

In this the main necessity was the settlement of an absolute unit of magnetic intensity (strength of the magnetic field, as we also say since Faraday's time), and of a method of carrying out trustworthy measurements in this unit. Both of these goals were reached by Gauss in a manner which is still of full value to-day. Unit magnetic field is defined by the fact that a unit pole is acted upon by unit force (one dyne) when placed in it; a multiplication of the pole strength and of that of the field both caused (according to Coulomb's law) a proportionate multiplication of the force. The method of measurement refers to the horizontal component of the field strength (in the sense of the parallelogram of forces), which is completely sufficient when the inclination is known.　Gauss' method for the absolute measurement of the horizontal component of the earth's magnetism, which cannot here be discussed in greater detail, goes far beyond its original purpose in importance, on account of the perfection with which it is developed, and has become of fundamental importance for all magnetic field measurements and hence also for the realisation of all common electrical units as used to-day.

The most admirable part of Gauss' structure is the perfect manner in which the conceptual difficulties which surround the 'magnetic pole' – the point of departure – are overcome, since first Gilbert and later Ampère, had made clear that such a pole is neither the seat of a peculiar magnetic fluid, nor in any way sharply defined in space.　The difficulty is overcome by the introduction of the 'magnetic moment' of the bar magnets used in carrying out the measurement, taking into consideration Ampère's results, already known at the time.　Since Ampère's theory of magnetism is still current to-day, Gauss' plan is still perfectly satisfactory.

Gauss was born of poor parents in a small and poor house in Brunswick.　His was a family of farmers; his grandfather had come into the town, where he earned his living as a gardener.　His father had many occupations, but finally also worked mostly as a gardener; he married Gauss' mother, who was the daughter of a stonemason, in a second marriage.　The family was very well brought up.　Young Gauss already showed signs at an early age of unusual gifts, even when at home and then at school; he learned reading by himself, only by asking how the letters were pronounced, and even before that his power of mental arithmetic, for which he had no preparation of any kind, astonished everyone. He also quickly mastered both ancient and modern languages. All this aroused the attention of a wider circle, even while he was at school, so that the fourteen-year-old boy was finally brought to Duke Karl Johann Ferdinand of Brunswick, who then became and remained Gauss' patron and protector.

Gauss must have already studied Newton's *Principia* when he was at school, and always praised it as his model; altogether and in every way, he retained a limitless respect for Newton. This early study allowed Gauss to begin his course at Göttingen University, where he remained only three years, by at once inventing the method of least squares, and this was followed uninterruptedly by further remarkable achievements, which belong to the most important advances in mathematics, but for the moment remained unpublished, with the exception of his famous dissertation, on the basis of which he was given his doctorate in Helmstedt without an examination. The circumstances of his industrious life were then fairly simple. Until the death of the Duke, Gauss received a fixed pension in Brunswick, so that he could devote himself entirely to work. In the year 1807 he then accepted a call to Göttingen, where he was professor of mathematics and director of the newly built observatory until he died at the age of seventy-eight. Gauss married twice. Both wives died young; but of his six children his youngest daughter remained to take care of him until his death.[1]

Gauss' own statements concerning his manner of thought and work would be instructive to the many people who, particularly in the school world, allow the high educative importance and effect of scientific investigation and science generally to be overlooked, instead of encouraging it, because they have not been taught better than to seek the main achievement in the mathematical and technical side, which is almost solely cultivated. As opposed to this, and perhaps more impressive than our remark already made relative to

[1] Gauss' family life is described in an account published in honour of the 150th anniversary of his birth, *C. F. Gauss und die Seinen*, by H. Mack, Brunswick, 1927.

Laplace's achievement, the consideration may be put forward that Gauss, the eminent mathematician, regarded analytical calculus – although he was a master of it – only as an appliance for carrying out without error work previously done by pure thought in the simplest manner, to such an extent, that he never sat down to put the calculation on paper, before he saw completely through the problem and regarded it as soluble.

What is now taught in all kinds of educational institutions in an increasing degree under the name of science is what Gauss – and of course everyone who is in the first place an investigator – regarded as a mere tool, a subsidiary matter; mathematical technique.   Physics taught in this way is not a field for the practice of healthy and simple thinking in direct contact with nature, whereby it could be of the highest value to humanity, but only a field for the practice of the technique of calculation, a pleasure for small minds.   In order that there may nevertheless be an appearance of actual science, experimenting in schools is increasingly overdone, technical achievements and showpieces are exhibited, generally of a particularly complicated description – again giving an opportunity for much calculation – while the simple ideas, which have led all great men of science by contact with nature to their successes and to an insight into the world, and which are also of educational value, are lost to view.

A consideration of the work of great men of science should help mankind to rid itself of this insane idea, to get out of its microcosm, back again to the macrocosm, to the greatness of nature itself, in face of which we should not however stand as masters, but as modest admirers.   The mastery of motors and wireless waves – by 'knowing all about them' – does not ennoble humanity; it coarsens and degrades them, and even makes them obviously more stupid.   On the other hand, the joy at newly found insight lifts them up, when understanding for it has been cultivated.

Joy of this kind was to Gauss actually the mainspring of his work; he says that he carried on his scientific undertakings only for their own sake; that is, from the innermost need of his soul.[1] This is also entirely in accord with the fact that he was never in a hurry to publish, since this was a secondary matter for him.   On the other hand he thus attained to an unshakable spiritual strength, which was connected with a deeply religious sense.   The foundation of it was a striving for truth and a feeling for justice.   He thus regarded the spiritual life of the whole universe as a great system of justice permeated with eternal truth, and this was the main source of his impregnable faith that the life of man does not end with death.

MICHAEL FARADAY
1791–1867

HERE again we have one of the greatest men of science, of his kind incomparable, and unique in the endowment with the peculiar mental constitution which makes the investigator, who uses his senses to penetrate into the unknown, and actually expects to see everywhere new things in plenty. He detects the smallest sign of anything peculiar or not fully understood, and then applies to it a mind already schooled by the thorough study of nature, and completely adapted to it, exhibiting the greatest patience in the pursuit and multiplication of his observations, until the new thing has been understood in its essentials, and thus has become a new mental possession.   The innumerable occasions on which Faraday brought these gifts to bear in his laboratory had to do with things of very varied importance; from discoveries of the greatest possible range down to small explanations,

[1] See the contemporary recollections of Sartorius von Waltershausen, *Gauss zum Gedächtnis*, Leipzig, 1856.

which are a necessity for the thinker, even if they do not always reveal anything essentially new. In this connection we see that even the most gifted investigator is unable to see ahead far into the unknown, and that his value rather consists in exerting his gifts faithfully and with the greatest industry, quite uninfluenced in his steady course by the desire for public success, purely from love of the depth of nature, and – as was the case with Faraday especially – simply from sheer delight in the observation of nature.

Only a small part of Faraday's whole lifework, which extended over all branches of non-living nature, can be touched upon here. His life from the time when he had found his place among men and had most fortunately attained quite early to freedom to exert his powers, was so completely devoted to science, that his life history and the history of his discoveries are necessarily one.[1]

Faraday was the third son of a blacksmith, who had migrated to London soon after his marriage, but came from the north of England (Yorkshire), where Faraday's grandfather was a mason. The family belonged to the small but strict religious sect known as the Sandemanians. After some teaching at school of the simplest description, Faraday was apprenticed at the age of thirteen to a small bookseller and bookbinder. At first he had to deliver newspapers, later, at the age of seventeen, he was allowed to learn bookbinding. Books on chemistry and electricity, which thus came into his hands, captured his attention, and led him to continual further study of books, and also to attending evening lectures of a popular scientific character, which were given in his neighbourhood. Finally, a customer of his master afforded him the possibility of attending several

[1] For a description of Faraday's life and work many biographies may be consulted (W. R. Burgess, J. C. Geikie, T. H. Gladstone, S. Martin, W. Jerrold, Silvanus P. Thompson). A recent one is by J. A. Crowther, 1918. Life and Letters, by H. B. Jones, and Faraday as a Discoverer, by W. Tyndall, both London, 1870, are the chief older works.

lectures at the Royal Institution. Thenceforth he developed the resolution to get out of business life, the manner of thought of which he hated, and devote himself entirely to science, which he loved.

At this time he wrote a request, in 1812, to the president of the Royal Society, asking for any scientific employment, no matter how menial, but this letter remained unanswered. With the aid of a book and practice on his own account, he had learned perspective drawing, and in the same way he made progress in all sorts of branches of learning by using the leisure time allowed him by his master.

How great was his progress both in manner of thought and in power of expression, is shown by very long letters, still preserved, which he exchanged with a somewhat younger friend who had had a school education. In the meantime he had written out neatly the lectures given by Davy, and illustrated them with drawings in a quarto note-book, and this he sent to Davy with the request to give him some kind of employment in his laboratory.

He quickly received a favourable answer. Davy immediately asked the advice of one of the governors of the Royal Institution as to what could be done with the petitioner. 'Let him wash bottles,' said the governor. 'If he is any good he will accept the work; if he refuses, he is not good for anything.' This course was followed, although Davy immediately saw that the young man appeared to be fit for something better. This decided Faraday's fate. He remained his whole working life in the Royal Institution, 1813–1858, first as Davy's assistant with a weekly wage, and finally as his successor.

No more suitable teacher of science could have been found for Faraday than Davy, who was undoubtedly more closely allied to him in mind than all other great men of science, as we see when we pass them in review. What a piece of good fortune that these two should have been contemporaries in the same place! If Count Rumford's foundation, the Royal Institution, had had no other success to show, the fact alone that it was a place suited to the work of Davy and Faraday would have been the finest possible justification for all time, of the foundation of the Institute, and its nature and constitution.

Faraday had already begun to experiment with simple appliances, so that he was immediately able to be an efficient help to Davy, in his preparation for experiments, and in the experiments themselves, among others also in the dangerous investigation of chloride of nitrogen then being undertaken by Davy. At the same time he was continually and enthusiastically engaged on his own self-education in every direction, for which purpose he joined a small society of kindred spirits, the 'City Philosophical Society,' the members of which strove to educate themselves by evening discussions and by holding, in turn, lectures. Then followed one of Davy's long continental tours, and Faraday was able to accompany him as his scientific assistant and also in a general capacity.

On his return, at the age of twenty-three, Faraday's independent scientific life began, although his position as assistant still remained for several years about the same. Davy first entrusted to him small independent chemical investigations, in which however he soon progressed to valuable discoveries, such as the first preparation of the compounds of chlorine with carbon, which he thoroughly investigated and analysed quantitatively. He was also busy on researches on sound, while at the same time he had to prepare the experimental lectures given at the Royal Institution, and to act as assistant during them. Here he was full of ideas concerning everything that could serve for the effective and worthy production of such lectures. He had also begun himself to hold lectures in the City Philosophical Society, which he delivered extempore, but nevertheless prepared carefully in writing, and this soon led to his remarkable mastery of description, and also of giving experimental demonstrations.

After he had thus tested his powers for about six years, and received Davy's approval, the apartments previously occupied by Davy in the Royal Institution building were promised him, and he married when thirty years of age the daughter of one of the Sandemanian elders, a goldsmith. The marriage was a happy one for the whole of his life, but no children came of it.

Ten years of enthusiastic and many-sided scientific activity now followed, of which we can only give a summary, since this time, as we see when we look back, only formed the introduction to his highest achievement, though its results would be quite sufficient to make a reputation for any man of science. To this period belong the researches carried out conjointly with Davy, on the liquefaction of gases. Substances originally discovered in the gaseous state, and hitherto only known in this state, such as chlorine, carbonic acid, ammonia, sulphurous acid, hydrochloric acid, were brought for the first time into the liquid state, a fact which at the same time confirmed the conviction of the tangible and ponderable material nature of gases, and of the unessential nature of the state of aggregation, which only depends upon temperature and pressure. Very extensive researches were also carried on concerning steel alloys, and the production of new kinds of glass.[1] Faraday further discovered at this time the hydrocarbon benzine, which has later become of so great importance.[2] He obtained benzine by thorough investigation of a product obtained from the condensation of oil gas, and most carefully determined its quantitative composition, vapour tension, and many other properties. This fundamental investigation deserves particular mention to-day,

[1] See *Faraday and his Metallurgical Researches*, by Sir Robert Hadfield, London, 1930; *Nature* (Supplement), Aug. 29, 1931.
[2] *Philosophical Transactions of the Royal Society*, 1825, page 440.

since the derivatives of benzine (the international name of which is benzol) form so important a part of chemistry.

These years were also signalised by Ampère's important work, which Faraday not only studied with great admiration, but immediately reproduced with his own and extended observations, whereby he exhibited the rotation of magnets about currents and of current-carrying conductors about poles, which is so important for the understanding of these phenomena. Though others had to some extent anticipated him in this matter, these studies nevertheless formed an effective prelude to the great discoveries he was soon to make, inasmuch as he became perfectly familiar with everything known concerning electromagnetism by his own personal observation.

Faraday was now forty years of age, he was an admired successor of Davy's in the delivery of the Royal Institution lectures, and also director of the laboratory, which thus gave him the most favourable possible circumstances for work. In the autumn of 1831 the discovery of electromagnetic induction occurred. This was an entirely new method of producing an electric current, at that time foreseen by no one, but which to-day, now carried out on the largest scale, has superseded almost all other methods. It has also made possible the production of very powerful electric currents, such as could never have been made by Volta's methods, and so laid the foundations of the magnificent success of modern electrical engineering.

The phenomenon could not have been predicted, inasmuch as it was not deducible from any known fact, but Faraday was an investigator who allowed innumerable analogies with the known, of the nearest and most distant description, to arise in his mind, and never became tired of doing this; he never regarded anything as impossible before he had thoroughly tested it, nor did he fail to use every effort to test it in every possible direction. Oersted and Ampère had obtained magnetism by means of electricity; might it not be possible to obtain electricity by means of magnetism? This was one of Faraday's simple queries. Furthermore, the phenomenon of influence is known, whereby an existing electric charge produces a fresh charge; might it not be possible to produce by means of an existing electric current a fresh electric current?

Guided by these ideas, Faraday prepared amongst other apparatus, a double coil of two insulated wires, which were wound alongside one another upon the same cylinder of wood. The first wire he connected to a voltaic battery, the second to a galvanometer. At first no motion of the galvanometer needle could be observed; the wire in which no source of current existed thus remained without a current, in spite of the presence of the current in the other wire. Even when a hundred and twenty cells were used for the first wire, no greater success was obtained. Nevertheless, Faraday was conscious of the richness of nature, which is revealed to him who knows how to approach sufficiently close to the unknown; it was his custom to pay attention to everything that could in any way be observed during an experiment. He noticed a slight deflection of the galvanometer needle, not during the flow of the current in one of the wires, but on the circuit being connected. Likewise, when the current was interrupted, a second deflection of the galvanometer needle occurred, this time in the opposite direction.

These phenomena were not of the kind expected, but they were there; the new discovery had been made. The battery current therefore 'induced' a current in the neighbouring wire, although not a permanent current such as is given by a voltaic battery, but only a short rush, such as is given by the discharge of a Leyden jar, and this rush of current occurred when the current was introduced into the first conductor, and was repeated in the opposite direction when the original current ceased. This fact, which depended upon a quite slight motion of the galvanometer needle, was thenceforward the starting point from which Faraday strove persistently to penetrate further into the unknown.

In the midst of work of the highest intensity, he then brought to light, in the course of three months, a large number of the most important new facts. First, induction by bringing together and separating the two conducting circuits, when no iron was present. Then we have magnetic induction, whereby Faraday first made use of a closed ring of soft iron, which carried two windings of wire on opposide sides; when the current was started and stopped in one of these windings, the needle of a galvanometer connected to the other was so strongly affected that it actually spun round several times. This was the first observation of a powerful inductive effect.[1]

He then used also iron rods with coils, and then he proved that simple steel magnets can also produce induction just as well as current-carrying coils, which fact entirely agrees with Ampère's conception of magnetism. The knowledge thus obtained enabled Faraday also to give a direct explanation of a group of phenomena discovered by Arago seven years previously, hitherto unexplained, and known by the name of rotation-magnetism. He then for the first time drew off steady currents by means of rubbing contacts from a copper disc, in which the current was generated by induction, by rotating the disc between the poles of a magnet. Finally he gave in the same first publication concerning induction, the general law of these phenomena, which summed up all the cases observed by him. It is a matter of the cutting of magnetic lines of force by conductors, and this entirely new kind of fundamental idea has also proved satisfactory in regard to all further observed forms of induction.

[1] A fine marble statue of Faraday lecturing with the ring in his hand, is on the staircase of the Royal Institution.

The magnetic lines of force were likewise first introduced by Faraday in this essential significance, and in one much wider in scope; they had hitherto been known only as the lines in which iron filings arrange themselves around a magnetic pole. For Faraday they soon became the most essential feature of the magnet, as also of electric current, as picturing the state in the space around the magnets and currents, and indeed everywhere where magnetic forces are present – everywhere as we say to-day, where a 'magnetic field' exists. Nothing has proved of greater importance and more fruitful than this conception of lines of force, as regards our continued progress up to to-day in this great domain of electromagnetic phenomena, opened up to us by Faraday.

At the same time the way was also pointed out to the use of induction phenomena, some decades later, as means for the production of electric current on a large scale; for every phenomenon, the laws of which are sufficiently well-known and reduced to quantitative expression, is also ready to be carried out on any scale desired. Faraday himself had thus shown the way from the weak galvanometer deflections which he observed to the dynamo machine of Siemens thirty years later; he was the discoverer of the natural processes from which modern electrical engineering derives its possibilities.

Already in January of the following year, Faraday had produced further new forms of induction; 'earth induction,' and 'unipolar induction.' The first, produced by the cutting of the earth's magnetic lines of force by moving conductors, became, in the hands of Wilhelm Weber, particularly important for the extension of our knowledge in this field, up to the foundation of the electrical system of measurement used to-day.

Faraday then allowed these investigations to drop for a time, so that the extraordinarily important discovery of 'self-induction' only occurred three years later. In the meantime he was busy with work on the identity of all kinds of electricity, in which he proved in detail that electricity from all known sources, including induction, is exactly alike in all its effects, which was necessary for mental satisfaction in view of the rapid accumulation of new facts.[1] In this connection, Faraday for the first time measured *quantities* of electricity by means of the galvanometer, whereby he proved, for example, that the quantity of electricity delivered by a certain number of revolutions of an electrical machine, when collected in a small or large capacity, and thus by higher or lower voltage, and then discharged through the galvanometer, always produces the same deflection. This is the principle of the measurement of electrical quantity by means of the 'ballistic' galvanometer, and at the same time also the proof that the current strength measured by the galvanometer actually represents quantities of electricity passing through it in unit time in accordance with the conception introduced by Ampère.

Further, in this period falls also Faraday's profound investigation of the chemical effects of electric currents, in which he made a further great step forward beyond Davy and Berzelius, by discovering, as a result of extensive quantitative investigations, the two laws still known by his name regarding the chemical effect of the current. In this connection he also introduced the terms generally used to-day – 'electrolysis,' 'electrolyte,' 'electrode,' 'anode,' 'kathode,' 'ion'; and he also already investigated the secondary chemical reactions so often occurring on the electrodes. On the first of these two laws he founded his voltameter, which measures current strength by means of the quantity of mixed gases produced by electrolysis. The second law, which connects chemical valency with ionic charge, has proved to be just as

[1] Davy had already interested himself in this problem, and had succeeded in showing the identity of the chemical effects of 'frictional' and 'voltaic' electricity.

Michael Faraday

WILHELM WEBER

fundamentally important as the researches of Dalton and Gay-Lussac concerning the combining weights and volume of chemical substances, and it further gave the first evidence of the existence of certain fundamental smallest quantities of electricity – the elementary electric charges – of which all larger quantities are composed, just as all material bodies are made up of atoms.

In the year 1837, Faraday developed the idea of electric lines of force, which represent the total effect of all electric forces, just as the magnetic lines of force do for magnetic forces. The course of the electric lines of force, from one electrified body through space to the other on which the opposite electricity produced by influence is seated, is not so easily visualised as the magnetic lines of force are, at least approximately, by means of iron filings. Faraday studied the course of lines of force experimentally, in the most admirable manner, by following out in detail the phenomena of influence, particularly in cases where the effect obviously takes place in curved lines around obstacles. Though this action in curved lines is by no means in contradiction with Coulomb's law, it is nevertheless usually only predictable in single cases with the greatest mathematical difficulty.

Faraday's conception of lines of force gives here, as in all other cases, a quick means of orientation, and it also goes beyond Coulomb's law and gives the essential of all electrical force effects, by a geometrical representation of the states of space, otherwise unknown, which condition these effects. It was a fundamental advance for all time, that Faraday here attempted to study the behaviour of nature by observation, from an entirely new standpoint. The gain appeared in its full extent, when thirty years after, Maxwell was able to express Faraday's conception of lines of force, which his contemporaries had not been able to follow, by means of equations.

However, Faraday's new point of view of electric force

Ss

already showed itself in the hands of its creator as immediately fruitful; he discovered the influence of the insulating material through which the lines of force pass upon their effect, a fact that had hitherto been unknown. Every insulator or 'dielectric,' is accordingly characterised by a peculiar 'dielectric constant,' which controls its application to the construction of large electric capacities. Here also Coulomb's law received a further extension, inasmuch as this constant appears in it as a factor, when the electric force no longer acts through empty space, but through material media.

After this succession of extraordinary discoveries, a period of exhaustion supervened for Faraday. These successes had not been obtained without very hard work; for though ideas came to him in plenty and without trouble, the necessary innumerable experiments, and continual modification of the ideas towards a convincing representation of actual facts – the true work of the man of science – had led to an expenditure of energy which reached the nearer to the limits of the possible, the further he was led by his ideas into the unknown.

Frequent excursions to the seaside, where he could only sit quietly and look into the distance, brought him some recuperation; but it was only when he finally undertook a journey to the Alps, and remained for a considerable time in and near Interlaken, which at that time was a quiet place, that he was completely restored to health. This was followed by a series of new discoveries and important works, in the time from his fifty-third to about his sixty-fifth year.

In this period he made his discovery of the magnetic rotation of the plane of polarisation. It is noteworthy that this discovery was preceded by hundreds of unsuccessful experiments with all kinds of crystals and other materials, without Faraday letting go of the idea of seeking some kind of effect of the magnetic lines of force upon light. 'It must be tried;

who knows what is possible ?' was one of his remarks; never-theless there was a deep conception behind it all, only one that he was not altogether inclined to trust; one and the same ether for the magnetic forces and for light. The first suc-cess came when he brought a piece of heavy flint glass, made in his earlier experiments of different kinds of glass, into the magnetic field. The success was a first sign of the correct-ness of the idea.

Also in the discovery which then followed of diamagnetism and the magnetic properties of all matter, this glass likewise brought the first success, since it was repelled out of the magnetic field, in a manner exactly opposite to the behaviour of iron and substances containing iron. Hereupon he made an investigation of many substances in the strongest mag-netic field that he was able to produce, whereby they were all shown to be either magnetic, or 'dia-magnetic,' like Fara-day's glass or bismuth.[1]

The course of the magnetic lines of force around currents and magnets, their inductive effects, and the measurement of magnetic field strength by means of these effects, occupied Faraday during the last period of his work. In this con-nection belong also ideas and experiments concerning a possible connection between gravitation and electricity, con-cerning also the rate of propagation of electric and magnetic forces, and a possible change in wave-length by a source of light (a flame fed with common salt), in the magnetic field. These latter matters led to no decision. However, his ques-tion whether the ether which carries light may not also be the medium for the electrical magnetic forces, was later answered affirmatively by Maxwell, who based his work upon Fara-day's achievements, and upon an important result found by Johann Weber. Maxwell's results were first a supposition

---

[1] It appeared that scattered observations concerning the repulsion of bismuth by magnetic poles had already been published, but no one pre-viously to Faraday had treated the subject as a whole and thus brought it into general importance.

based on his equations, and this prepared the way for
H. Hertz to confirm them experimentally and thus discover
electromagnetic waves. The idea likewise of an influence
of magnetic field upon light sources turned out later, when
means of observation had been very much refined, to be
correct,[1] and became of no less importance for all further
developments of our knowledge right up to the present day.

In the year 1858, Faraday left the Royal Institution and
went to live in a house outside London presented to him by
the Queen. He began to suffer from rapidly increasing loss
of memory, but was nevertheless able to enjoy another ten
years of life in his own modest fashion. He died at the age
of seventy-six.

Faraday combined in a rare manner great pride with the
highest modesty, and this peculiarity of character depended
entirely upon a deeply religious foundation, which however
he kept concealed. He remained always true to the small
sect of the Sandemanians, which at that time also was
scarcely known to the outside world, but in which he was
born, and the fundamental idea of which was a simple
adherence to Christ's teaching, which was to be interpreted
by lay preachers from among the brethren. How deeply he
entered into everything that came into the range of his
thoughts, may be seen from his definition of friendship,
which he gave in one of his letters, in between an account of
chemical experiments, at the time when he was a bookbinder's apprentice. 'A friend is he,' he says, 'whom one is
ready to serve next to one's God.' This serious point of
view he retained in every relation, both towards men and
towards science, throughout his whole life. Disappoint-
ments in regard to human relations could not fail to occur,
and Davy already prepared him for such, when he smiled as
Faraday once remarked to him that he thought that men of
science are animated by higher moral feelings than other men.

[1] 'Zeemann effect.'

JOHN TYNDALL

JULIUS ROBERT MAYER

JAMES PRESCOTT JOULE

Faraday knew how to avoid the difficulties which he would have met with, living as he did upon a high and lonely level of thought and feeling, which he was unwilling to leave, among the average of mankind; he avoided all his life activities and memberships to which he, as he was, would not have been adapted. For this reason he believed himself obliged to refuse, in the most definite manner possible, the presidency of the Royal Society which was offered to him,[1] and he likewise refused to accept a title. He wished to remain simple Michael Faraday, but this did not prevent the Court treating him with great honour. It is certainly no good sign of the nature of the world, that Faraday had to regard himself as unsuitable for such relationships. Newton, who was not by any means more of a man of the world, or more shallow a character, did not regard this as necessary in his time, although he was not without his embarrassments. However, Faraday's desire for the greatest possible retirement agreed, as did also his rejection of profitable activities, entirely with his intention of remaining completely devoted to science. In this respect the laboratory of the Royal Institution was ideally adapted for his work; his only assistant there was an old soldier.

Along with this, he gave, in the large lecture theatre of that Institution, his experimental lectures for an audience drawn from all circles – being in this respect also Davy's great successor – and at Christmas he always added a few special lectures for children.[2] Thus Faraday, in spite of his retired mode of life, was a personality well known and greatly honoured by all circles in London, and his

[1] This refusal had for Faraday in later years a curious consequence, which would certainly not have occurred under his presidency; one of his papers was refused by the Royal Society. It was then only published together with his posthumous papers (*Life and Letters*, vol. 2, pages 411–418).

[2] Of these lectures, one in particular later appeared as an immortal example: *The Chemical History of a Candle*; see the edition with preface by J. Arthur Thompson, London and Toronto, 1920.

discoveries also brought him much recognition from distant parts. All such signs of goodwill which fell to him he accepted with the greatest warmth and at the same time modesty; but it would nevertheless be a mistake to suppose that his fiery nature and his feeling for a just estimate of humanity did not become evident when he felt that he was in contact with worthless people. Thus he says for example, in a confidential letter (written in 1853) that he prefers the obedience, faithfulness, and instinct of a dog by far to the average stupidity of mankind.

In many respects he united in himself the most strongly marked qualities, which might appear to the un-understanding as opposed to one another, but this was the distinguishing characteristic of his rich nature. Thus he was, as he once himself affirmed, always very much inclined from his youth upwards to regard the most incredible things as acceptable; but a single fact determined by observation always sufficed to destroy a structure in thought, no matter how beautiful; and however much pleasure he had had in erecting such structure, his final and unconditional guide in every respect was truth, agreement with reality. In one of his experimental investigations he thus describes his ideal of the man of science; enthusiastic, but careful, linking experiment with analogies, mistrustful of preconceived ideas, regarding a fact as more valuable than a theory, not hasty in generalisation, and above all prepared to test his own opinions at every step afresh both by consideration and observation.[1] He himself fulfilled this ideal in the fullest possible manner, and his character as a man of science was his character altogether and in everything; great minds have never been divided against themselves.

He wished to have only his relatives and his nearest

[1] *Experimental Researches in Electricity*, London, 1839, vol 1, page 360; No. 1161. These have been issued as cheap reprints in the Everyman Library.

friends attend his funeral, and his grave was to be of the commonest description and in the most simple position, which also was followed out.

WILHELM WEBER

1804–1890

WEBER was the founder of the electrical system of measurement generally used to-day, and he became so by being the first to work out thoroughly and with the greatest possible refinement the newly discovered domain of knowledge opened up from Oersted to Faraday, and otherwise linking up with what Gauss had already done as regards magnetic quantities. For this purpose he devised new and much improved appliances of different kinds, such as the electro-dynamometer and the earth-inductor, and carried out with the greatest energy measurements of an accuracy hitherto unobtained, which measurements brought all the new discoveries into so firm a connection with one another, that no gap remained. The result was that a beginning could at once be made to operate with quantities the conception of which had only just been formulated, or at any rate made measurable, such as electrical current strength, induced electro-motive force and electrical capacity, just as had been possible since Newton's time as regards forces, velocities and masses.

In the course of this work Weber himself also made a new discovery, which could not fail to be the case with any thorough and devoted investigator. He found, when he brought into connection the two laws of Coulomb for the electric and magnetic forces – which was possible by using the connection discovered by Oersted – that in this connection a velocity played a part, which he determined by means of a difficult experimental investigation, and found to be identical with the velocity of light as known since Roemer's time. Here for the first time the velocity of light was found to play a decisive part in the domain of purely electromagnetic phenomena, the velocity which is so characteristic a magnitude for ether waves, that is, light. It was legitimate to regard this as a definite sign of the correctness of Faraday's idea, that the electromagnetic forces are states in space which have to do with the same ether as that which carries light. This idea was later thoroughly worked out by Maxwell, whose work rested in particular upon Weber's researches.

The velocity of light had also appeared in another connection – also within the range of Weber's researches – as a determining electrical quantity, namely as the velocity of propagation of electric tension along telegraph wires. The electromagnetic telegraph had been introduced by Gauss and Weber in the year 1833,[1] and had very soon spread further, and this led to the velocity being measured by means of a rotating mirror; it was found to be practically equal to the velocity of light, which fact also became of importance as regards conclusions to be drawn later.

Weber was also the first who attempted to develop quite

---

[1] Telegraphy by means of light signals had already long been in existence. Ideas of, and attempts at, electric telegraphy were also fairly old; very shortly after Gray had discovered the possibility of conducting electricity to a distance, a beginning was made with telegraphy, and every newly discovered electrical effect – in particular the chemical effects of the current – gave a new spur to plans of this kind, but in every case it was supposed to be necessary to use almost as many wires as there are letters in the alphabet, and hence no practical results were obtained. Gauss and Weber were certainly the first to carry out electric telegraphy with two wires only, for they signalled to one another between the observatory and the physics institute in Göttingen by means of the deflection of a magnet, the current being produced by induction, so that no other source of current was necessary. Very shortly afterwards (first in Bavaria), it was found out how to work with only one wire, by using the earth as return for the current. Gauss and Weber absolutely refused to attempt to gain any personal advantages for themselves from the introduction of the electric telegraph.

generally the idea of definite smallest amounts of electricity – elementary electrical quantities – which was already given by Faraday's second law of electrolysis, and Davy's and Berzelius' ideas concerning chemical forces. He ascribed to these particles for the first time not only a definite charge, but also a definite mass (inertia), and developed in all essential points the conceptions of the conduction of electricity in metals which were reintroduced more than thirty years later, when they were suggested by the study of cathode rays and the further results of this study. In this connection, also, Weber already gave an explanation of diamagnetism.

Not a few other results of Weber's long and industrious life cannot here be touched upon, and in the case also of other recent investigators, with whom we are to deal, we shall likewise have to reserve the forefront of our description for that which made the older men of science great and admirable: the discovery or realisation of the fundamentally new. Weber still belongs to the investigators who are well able to bear the application of this measure, and if much has to be passed over in his case and in the case of other more recent men of science, this is only due to the fact that the old investigators left behind so much that was worthy to be further worked upon by the finest minds, but at the same time has so well stood the test of this that nothing essentially new has appeared in the process.

Weber was the fifth child of a theologian in Wittenberg, whose father had been a farmer. He studied science in Halle, and there became a member of the university, and afterwards assistant professor. In the year 1831, he was called to Göttingen at Gauss' suggestion, to the chair of Physics. He there became collaborator with Gauss in the latter's magnetic investigations. In the period of collaboration with Gauss came an event which had a strong influence on Weber's life; he belonged to the famous Göttingen Seven. King Ernst August of Hanover in the year 1837 abolished by

his own autocratic act the parliamentary constitution of
the State, though he had confirmed it by oath, and against
this act Weber and six of his colleagues in Göttingen
issued a manifesto, the consequence of which was their
dismissal.[1]

Weber was now five years without a post or a fixed income. A collection was made in the whole of Germany for
the benefit of the Seven, from which Weber received 1400
Talers.[2] Weber however did not feel that he should make
use of this gift, but put it into safe keeping, and lived in an
extremely modest manner in a small room. The apparatus
needed for his work, which at first was not very much, he
received from Gauss.

This work related in the first instance to settling an absolute unit of current strength. Weber based this unit on the
magnetic effects of the current, according to which unit
current would be that current, unit length of which would
exert unit force, at unit distance, at right angles, upon unit
magnetic pole as defined by Gauss. This idea was realised
by means of the tangent galvanometer, for which purpose,
however, Weber was obliged to refine the instrument, and

[1] Weber certainly contributed to the feeling against the Prince by
remarks which he made at table in the presence of W. von Humboldt:
that German professors had no native country at all, and were no better
than dancers, who were ready to go anywhere where they were offered
a few pence more. The brothers Grimm were members of the Seven,
but not Gauss. The latter fact is readily understood, since Gauss had
extremely little opinion of 'constitutional governments with decision by
majorities,' for he had a very low opinion of the understanding and
morality of the great majority, and thus regarded revolutionaries always
with great mistrust (see *Gauss*, by Sartorius von Waltershausen, page
94). Gauss no doubt had adopted an unusual attitude by holding this
view; but he also acted quite obviously, just as did the Seven, in accordance with his convictions. Weber's view must have agreed with that of
Jacob Grimm (see the latter's statement in *Kleineren Schriften*, edition
of 1911, Berlin, pp. 28 ff.). Where should we find to-day even a few
professors, like Gauss or the Seven of Göttingen, who would make their
decision from any other point of view than, at the best, the needs of their
faculty? ('Professor' originally means a person who *professes* a certain
faith.)

[2] This was almost double his yearly salary.

the method of calculating its results, in a very high degree.[1]

In order to fix the unit current thus realised for all time in a simple manner, Weber, relying on Faraday's first law of electrolysis, made use of the voltameter. The choice of the unit current (now called the ampère) and also the means of fixing it, are still valid to-day.[2]

In the year 1843 Weber was called to the University of Leipzig, where he then invented his electro-dynamometer, which depended upon Ampère's discovery of the force exerted by one current upon another, and was suited for the finest and most accurate measurement of these effects.

Six years later Weber again returned to Göttingen, where he remained for the rest of his life. He then began the investigation which led to the settlement of the absolute unit of electrical tension or electromotive force (called to-day the volt), for which Weber relied upon Faraday's law of induction, which was possible without the use of the idea of lines of force, since he made use of Gauss' measurement of the earth's magnetic field and the induction produced by it. His earth inductor, with which he made very extensive quantitative measurements, afterwards became one of the most important appliances in electromagnetic measurement.

[1] This also allowed a first exact proof of the law, according to which the effect of current of any length and form upon magnetic poles in any position can be calculated. This law is usually given under the names of Biot and Savart. These scientists only investigated rectilinear currents (see Biot's *Elements of Electricity and Magnetism*, trans. J. Farrar, Cambridge, N.E., 1826); Ampère took for comparison zig-zag currents, and an observation by Laplace gave the generalisation for 'current elements' (short lengths of current), in any position as compared with the magnetic pole. The circular currents in the tangent galvanometer then gave the best case for an exact test.

[2] The realisation of this in a later and more refined manner, which reached an exactitude not exceeded to-day, was performed in the year 1881 (and repeated in 1886), by Friedrich Kohlrausch (1840–1910, president of the Reichsanstalt in Berlin); he measured the exact amount of silver deposited by one ampère flowing for one second.

From the units of current and voltage thus settled, Weber was able, by means of Ohm's law, at once to deduce the absolute unit of resistance (to-day called the ohm), which is particularly suited for realisation and preservation in the form of wire, or much better still, in the form of a thread of mercury, as used by Siemens since 1860 (Siemens unit); and hence a large number of Weber's latest measurements related to this unit. Since the means required for this purpose continually increased, but the State believed it necessary to economise, Weber made use, for the first time, of the money subscribed for him when he was dismissed from office. Weber was not able to witness the complete realisation of all the refinements which he had planned for settling the electrical units, in spite of the high age of eighty-six which he reached; but what he began was continually pursued, most of all in Germany, and resulted in our present-day certainty in electrical measurement, which fulfils every possible requirement.

During Weber's lifetime, in the year 1881, there was a Congress in Paris, which arranged the international acceptance of the system of units founded by Gauss and Weber. Here the main question was the settlement of factors by which the absolute units were to be multiplied in order to provide practical units of a suitable size, and also to fix short names for the units. The numerical factors were necessary, since all the units had been derived from the three fundamental units of space, mass and time, solely by means of natural laws and without any reference to their size. Thus for example, the absolute unit of electromotive force would have been much too small for ordinary use, and it was decided to use a unit 100,000,000 times as large, and call it the 'volt.' For all this, and also for the settlement of the fundamental units (whether, for example, the centimetre or the millimetre was to be taken as unit of length) an agreement between the various countries was necessary, and such an

HERMANN HELMHOLTZ

RUDOLF CLAUSIUS

LORD KELVIN

agreement can never take place without a certain amount of arbitrary decision.[1]

It is a strange fact that in choosing names for the units, the names of the originators of the whole system of units – Gauss and Weber – were entirely unused, and surprise was immediately expressed at the time.[2] It is natural that in Gauss and Weber's own country these objections have been particularly emphasised, and have not yet been silenced.[3]

Weber was extremely simple in his tastes, with a childlike happiness of temperament and very contented by nature, but yet in his way of thought immovably upright and of the strongest character, and this now and then even allowed him to exhibit a certain sharpness of temper, which he also showed in public when he felt it to be necessary, although he in general avoided such actions. His trust in human nature easily led him much too far, and this was made use

[1] Helmholtz's was the principal representative of Germany at this Congress; his influence was the greater because he showed great readiness to meet the wishes of others present in Paris. Weber was invited to the Congress, but his refusal was almost self-evident in view of his age of seventy-seven. The Congress was reported in *Nature*, September 6th, 1883.

[2] See for example, B. Wüllner in his *Lehrbuch der Physik*, 4th Ed., 1886, vol. iv, p. 922, and Werner von Siemens, *Lebenserinnerungen*, 4th Ed., 1897, p. 281. The excuse was made for the omission of Weber's name that a 'Weber's unit of current' was already in use, which was ten times smaller than the one chosen by the Congress, and hence might lead to confusion. This excuse does not hold water; for in the first place the said smaller unit, though known in Germany as Weber's unit, was nowhere in actual use (current measuring instruments with fixed graduations in units did not exist), and in the second place the larger unit chosen by the Congress was actually in use under the name Weber in England, where the technical employment of electric currents was at that time further advanced. The fact that Weber still lived, whereas Ampère, whose name was used instead of Weber's, was already dead, has curiously enough never been put forward as a reason, though at the time it might have been, though unmentioned, one of the grounds for the decision.

[3] A recent practice has been to use Weber instead of Ampère as a description of the current unit at will, which cannot produce any confusion. Gauss has likewise been used for some time (without the decision of any Congress) to designate the absolute unit of magnetic field strength.

of by the spiritualists, who were particularly active in his day.[1]

His kindness and friendliness did not fail him even when these good qualities resulted in his being pushed into the background; accordingly no sign of offence on his part over the behaviour of the Paris Congress on units, not even any remark of his on the subject, has been reported.[2] Weber always took a great interest in the fate of the German nation and all events connected with it; thus the foundation of the united Empire in 1870 was hailed by him with great joy. He was small of stature, and did not look very robust; but he was an energetic and tireless walker up to his old age. He always remained in close contact with his brothers; he was never married; a niece kept house for him. He died at the age of eighty-six.[3]

[1] Weber was not alone by any means in this respect; a whole circle of older professors, whose eyes and ears were not adequate to cope with conjuring tricks, and whose knowledge of human nature was equally defective, were also made victims. In England also, generally speaking, the same thing happened, as is shown by the example of Crookes; Faraday however repeatedly declined the invitations of the table-turning spirits in the most decided manner, when they were issued to him. It is to-day much easier, after so many exposures and confessions, to judge this misuse of science than it was at the time. [The English reader may be referred to Podmore's *The Newer Spiritualism*, London, 1910; but of course the literature of this subject, in all shades from extreme faith to extreme doubt, is endless.]

[2] On account of Weber's ripe age at the time, the want of recognition shown was easier to bear, since it could no longer act injuriously upon his facilities for work.

[3] The reader is referred to a sketch of his life by his nephew Heinrich Weber (Stuttgart, 1892); whom I also have to thank for valuable information by letter.

JULIUS ROBERT MAYER (*1814–1878*)
JAMES PRESCOTT JOULE (*1818–1889*)
HERMANN HELMHOLTZ (*1821–1894*)

JULIUS ROBERT MAYER, a modest practising physician in the town, at that time very small, of Heilbronn on the Neckar, was the discoverer of, and first to announce, the principle of the conservation of energy. He perceived that, beside the laws of motion of Newton and Galileo, another general law holds throughout nature, the grasp of which meant throwing light on just those parts of science which had hitherto remained dark in spite of investigation, and indeed meant the linking together of several separate branches of knowledge which had hitherto appeared disconnected and indeed contradictory. At the present day, the energy principle may even be described as the highest of all natural laws, inasmuch as it holds for all phenomena of matter as well as of ether, and further, no limits of any kind to its validity have showed the slightest signs of appearing in all the eighty years during which it has been known and applied.

Mayer announced his new discovery in very few words in the year 1842, in Liebig's *Annalen*, a widely read scientific periodical, in which words the essential point, the calculation of the mechanical equivalent of heat, is already stated; it was later discussed in detail in a separate paper which appeared in 1845. He had a heavy penalty to pay for all this. It is necessary to go back to Galileo to find an adequate analogy to his suffering on account of a new and profound idea. But since no two individual cases ever show exactly the same course, here also a difference is present: Galileo was opposed by dark powers hostile to natural science; Mayer's fate lay within science itself. The great majority who were the official representatives of science did not understand him, the few who did, who immediately grasped the idea, did not like him.

Matters went so far that when the new ideas finally began

to be accepted, Mayer found that they were actually passing entirely under other names, not merely abroad, but even in Germany. This also finally extended to the daily newspapers, so that Mayer, who was a very popular doctor in his birthplace, soon began to be looked at askance there as someone who pretends to have laid claim to the discoveries of others. After having been repeatedly repulsed by the editors of German learned periodicals, and after having also written to the Paris Academy without success, he turned to a German daily newspaper, which was widely read in learned circles, and which he regarded as of sufficiently good standing to serve as medium for a public explanation and for making his discovery known under his own name. His contribution was also accepted; but soon after there appeared in the same newspaper an article which, written apparently from a competent writer, made Mayer and his discovery appear ridiculous and unscientific.[1]

Mayer now repeatedly applied to this newspaper, which was owned by a publishing house of Stuttgart and Tübingen of high reputation, to insert a suitable correction; but without success.

In this way, having been tortured for over five years by similar experiences, Mayer became more and more nervously excited; finally, on a sunny May morning in the year 1850 he suddenly sprang out of the window, after a sleepless night. The window was thirty feet above the ground and

---

[1] The article was signed by a doctor of philosophy. The same doctor was soon afterwards received into the university of Tübingen as *Privatdozent*. Another *Privatdozent*, on the other hand, Eugen Dühring in Berlin, who had thoroughly emphasised Mayer's achievements for the first time in historical connection in an article called 'Critical History of the Principles of Mechanics' (1872), which had been given a prize on the decision of Wilhelm Weber, was driven out of the university, when in the second edition of this work (1876) he added some strong expressions, and in another publication exposed further scandal in the university. All this shows how little was thought of Robert Mayer, both in his own home university, as also in the capital, where Helmholtz was then representing physics.

Mayer injured his legs badly, and only gradually recovered in Wildbad. This was by no means the end of his troubles. They even became worse, in a manner which did great discredit both to the representatives of science as to those of medicine. But we will proceed with this story after first discussing his work.

When we read to-day Robert Mayer's three main publications[1] we can only once more be astonished at the richness of their contents; they bring in the main already everything fundamental to the energy principle, and we may also be surprised to find therein very much that, even to-day, is still fairly generally ascribed more to his successors than to himself.[2] The first paper (1842) puts the discovery in a very condensed form, without any argument based on definite facts;[3] however, the mode of calculating he tmechanical

[1] 'Bemerkungen über die Kräfte der unbelebten Natur,' 1842; 'Die organische Bewegung in Zusammenhang mit dem Stoffwechsel,' 1845; 'Beiträge zur Dynamik des Himmels,' 1848. A translation of these will be found in *The Correlation and Conservation of Forces*, a series of expositions by Grove, Helmholtz, Mayer, Faraday, Liebig, and Carpenter, with introduction and notes by Youmans; New York, 1865.

[2] Thus for example, the idea which has become more and more important in regard to the temperatures of the fixed stars, that shrinkage under the influence of force must result in the production of fresh quantities of heat, is already clearly stated in the paper of 1845 (second part of section 3). Later (in the 1848 paper) Mayer put forward the fall of meteoric masses into the sun as the chief means of maintenance of its temperature, but without at all cancelling the other suggestion.

[3] The turns of expression used there and elsewhere by Mayer, such as 'ex nihilo nil fit,' 'nil fit ad nihilum,' 'causa aequat effectum' (nothing is made out of nothing, nothing is annihilated, the cause is equal to the effect) are not to be regarded as proofs or preliminary axioms (as has often been stated by way of adverse criticism) but as short expressions of our knowledge, thrown off by way of humorous explanation. Mayer was also fond at other times of these turns of speech, and his humour was well known. These phrases may be regarded – following Mach – as the sign 'of a mighty, instinctive, and still unsatisfied need for a substantial view of what we to-day call energy,' in which mode of expression 'Mayer in no way behaved differently from Galileo, Black, Faraday, and other great men of science.' The view of energy as a substance has actually, quite recently, turned out to be perfectly tenable (see the remarks concerning Hasenöhrl, at the close of the book). The humour of great minds is always of significance; it is always understood only long afterwards.

Ts

equivalent of heat, and a sufficiently approximate value of it, are given. The second publication (1845) states everything at full length, and in conclusion adduces many applications to living nature; the third publication applies the energy principle to celestial processes.

The contents of these papers show that Mayer was the first who (already in the year 1842) saw clearly and in the most general way what had gradually been come upon by a number of investigators, in the course of the development of knowledge either in fragments or in the form of questions of fundamental importance, but was still without foundation. We already find in Galileo, Huygens, and Leibniz, questions regarding mechanical work performed by gravity and other forces, and Huygens was the first to recognise the importance of the product of mass into the square of velocity in phenomena of motion, for example in elastic collision, where the sum of these products remains unchanged, however complicated the phenomena of motion. The product is now of course known as kinetic energy, when divided by two.

Then came the great difficulties of the question 'what is heat?' which had been illuminated by the experiments of Rumford and Davy, in which heat had appeared without having anywhere disappeared, whereas its quantity had otherwise – in particular in calorimetric measurement – proved to be just as invariable as a quantity of any kind of matter. When the phenomena of the electric current came to be known, the heat of the electric arc and of a wire made to glow by a current again presented a riddle; for no compensating cold appeared, excepting only in the case of thermo-elements discovered by Peltier in 1834. As regards Volta's elements, Faraday, as well as Carnot before him, had remarked that they exhaust themselves in the production of all phenomena of whatever kind, and the cause of this might be supposed to be due to the chemical changes shown by Faraday to accompany invariably the passage of current to the cells. All this and a great deal more, which had either remained questionable or merely isolated fact, was suddenly brought by Mayer, in the most astonishing manner, into a grand connected system. He notices the fundamental importance of the nature of heat, and answers this question by saying: 'Heat is a force; it can be transformed into mechanical effect.'[1]

He there sees how heat is transformed in the steam engine into mechanical work, so that a certain number of calories disappear, while a certain other number of metre-kilograms of work appear. He also sees how that product – mass multiplied by square of velocity – likewise denotes a quantity of work; for velocity only results from the expenditure of work, and when the product decreases, some form of work done appears, such as the raising of a body thrown upwards, or in the pendulum; and heat also may be this form of work, as when a moving body comes to rest by friction, or in the case of inelastic collision, where the product also diminishes, while heat is produced, which is not the case in elastic collision.

He further sees that likewise the mechanical separation of

[1] We say to-day, in the same sense that Mayer used the word 'force,' or 'effect,' rather 'energy,' which expression was introduced much later (Helmholtz still used the expression 'conservation of force'). 'Energy' is stored-up work, and since work is measured by the product of the working force and the distance moved in the direction of this force, work is thus something other than force defined in Newton's sense (as Leonardo already recognised); it is therefore not permissible to use the word *force* in place of *work*. Mayer was perfectly clear on his point (as his publication *Remarks on the Mechanical Equivalent of Heat* in 1851 showed in full detail); only he thought that it would be possible to give up Newton's definition of force. This latter fortunately did not happen; instead, it became clear, that Mayer's newly formed concept of stored-up work – in which work appears as a substance unalterable in amount – also needed a new name, and *energy* was introduced for this purpose. This also again shows us the great novelty of Mayer's achievement; its formation of a new concept required the introduction of a new word. The fact that the concept arrived before the word is a sign of the solidity of scientific investigation at that time, quite by way of contrast in many other cases, where the opposite process so slyly recommended by Mephistopheles, takes place: 'for just where ideas are wanting, a new word appears at the right moment' (*Faust* I, Fourth Scene).

bodies which attract one another, such as the raising of a weight, signifies a storage of work, and that in the chemical separation of bodies the same thing holds; 'the chemically-separated existence' for example of carbon and oxygen or chlorine and hydrogen is also 'a force' ('form of energy' as we prefer to say to-day), and when this form of energy disappears, namely when the separation ceases, heat is again produced, as in a coal-fire. Thus Mayer for the first time throws entirely new light on the old and well-known production of heat in chemical processes, which had first been regarded as the setting free of a chemical component (heat-stuff), and then – when this explanation was shown to be incorrect – remained devoid of all explanation.

This new point of view put the whole matter and many others in a new and important connection. It thus also becomes clear that the development of heat by electric currents takes place at the cost of the chemical energy used up in the voltaic cells. Also, the work stored in electrically charged bodies, which is shown by phenomena of attraction and repulsion, in the effects of the spark and other heat phenomena on discharge, can only be produced by expenditure of work – as Robert Mayer made clear for the first time – for in every such case it is a matter of separating the two opposite kinds of electricity, and since these attract one another this separation can only be effected by expenditure of work, whether this takes place in the electrophorus by influence, or in the frictional electric machine, or in Volta's manner by contact between conductors of the first class.

Mayer thus arrives at a statement of the five different forms of energy, which we also enumerate to-day: 1. Potential Energy (energy of position); 2. Kinetic Energy (energy of motion); 3. Heat; 4. Electro-magnetic Energy; 5. Chemical Energy.[1] 'In all physical and chemical processes the given

[1] Mayer's terms are only slightly different (see the previous footnote), and he explained quite unambiguously the terms which he uses.

force (energy) is a constant quantity.' Only the form of
energy is changed; its quantity remains unchanged. 'There
is in truth only one single force (energy). In eternal change
this circulates in dead as in living nature. In both no process takes place without change of form of the force (energy).'
Mayer then gives twenty-five pertinent examples of energy
transformations, or changes of position of energy, the quantity remaining the same, which, after what has been just said
with equal impressiveness concerning the different forms of
energy, cannot leave the slightest doubt in anyone who has
even glanced through this essay and is sufficiently prepared
by his own thought, concerning the sense and extent of
Mayer's achievement.

Mayer continues with an extensive discussion on energy
of relationship in living nature, in which he as a doctor took
a particular interest.[1] Here again he starts from the highest
point of view, beginning with the energy of the sun, which
pours upon the earth in the form of light, and here, besides
warmth and raising of water to the clouds, also produces the
chemical energy of plants, and hence that of animals. It is
also important that he clearly and distinctly points out that
in muscular work there is a direct transformation of the
chemical energy of food into mechanical (potential or
kinetic) energy, without the intermediary of heat energy,
which is required by the steam engine.

It is obvious that this magnificent structure of ideas,
which this simple doctor had developed in the course of a
number of years, at first could only be hypothesis or supposition, and could only gradually become transformed into
tried and tested knowledge after many quantitative comparisons had been made with reality. This transformation –
to which Joule at once greatly contributed – has been completed to-day, and has not changed anything of importance

---

[1] The title of the essay already puts this part in the forefront, after the
first paper had been limited to forces of non-living nature.

in Mayer's edifice.[1] Nevertheless, the system set up by Mayer gave from the start the impression of being true to nature; for it threw light upon so many processes in nature that had hitherto been completely in darkness; it provided such a wide-ranging summary of knowledge, as would not have been possible if any of its main points were not correct. Just as in the case of Newton's structure of ideas in the *Principia*, where all the knowledge of nature collected by his most eminent predecessors provided proofs in abundance for his system, so also in the case of Mayer's system of ideas, where the proof was provided by the various investigations we have mentioned.

However, Mayer himself had already undertaken a comparison with reality in certain special cases, and here he proceeded quantitatively. He calculated the number of metre kilograms (potential energy) resulting from the transformation of one calorie of heat – in other words the mechanical equivalent of heat – by using existing measurements of the specific heat of air at constant pressure and constant volume in an entirely new manner. The fact, grasped by Dalton and confirmed by Gay-Lussac and Laplace, that air is cooled when it expands against external pressure, is regarded by Mayer as a case in which the heat which disappears has been turned into work, which is done against external pressure, and he based the admissibility of this view upon the fact that a gas when expanding into a vacuum – that is to say without doing work – does not show any cooling, as had already been proved by Gay-Lussac in the year 1807. So that, in fact, the performance of work and consumption of heat are connected together in the closest possible way.[2]

---

[1] We might mention as a necessary correction made afterwards the introduction of the factor $\frac{1}{2}$ into the kinetic energy.

[2] Among many statements brought forward against Mayer's view, and one which was only with difficulty and after some time cleared up, since it was brought forward by persons of apparent authority, is the statement that his calculation of the mechanical equivalent of heat is not free from

Since figures were already known for the quantity of heat in question, and also for the change in volume at given pressure, Mayer was able to calculate the equivalent by means of an imaginary experiment, no actual fresh experiment being necessary. Using the figures of that day he came fairly close to our more accurate figure of to-day, 427 metre-kilograms for one calorie. He also remarks that other gases, according to the measurement, give the same value as air, and that this is a confirmation of his view. Furthermore, he himself arranged an experiment in which work was transformed into heat, by measuring in a paper factory the rise in temperature of the pulp, which was being stirred with the expenditure of five horsepower; the resulting amount of heat was found, when the losses were taken into account, to agree with his equivalent.[1]

We see that Mayer produced altogether quite enough experimental basis. Apart from anything else, the experience that perpetual motion was impossible, so often used for the construction of important conclusions, comes in here with its full weight; indeed, it occupies the central position of the whole; it furnishes the assurance that an increase of energy cannot take place in any known process. The apparent fact, long regarded as true, that a *loss* of energy can actually take place, as in the case of every system of frictional processes, was proved by Mayer to be a delusion, for he showed that in such cases there is a transformation into heat, which is likewise a form of energy. With this, the complete invariability of the total sum of energy was made clear, and hence it was above all necessary to clear up the processes connected with heat, in order to arrive at an idea of the way in which energy is concerned. This clarification was also

objection. This statement ignored Gay-Lussac's discovery above-mentioned, and Mayer's express reference to it.

[1] Mayer expressly stated that more exact experiments were needed; but for his own person he regarded the continuance of his medical practice as the more necessary.

effected by Mayer, and quantitatively, by his calculation of
the mechanical equivalent of heat. All this was overlooked by those who afterwards further developed Mayer's
ideas, and by those critics who did not give Mayer his due.
As often happens when an entrance door has been opened,
entry then appears as something quite natural, as though the
door had never been closed, and as though its discovery and
opening did not form the main achievement, after which all
further steps become almost a pastime.

Why was it reserved for Mayer to find the entry, to recognise and to open it, and even for himself to use it to survey
fairly thoroughly the new territory ? Why should he alone
(and, though in a hidden and unpublished form, Carnot, who
died young) have been chosen for this task, so that all others
could only follow, although they may have had similar
ideas, but had not brought them to this degree of perfection ?
Without any doubt, it must have been the peculiar nature of
Mayer's mind which enabled him to effect this, and it must
have been a very rare type of mind; for the ideas on which
the whole thing depended had been in the air, so to speak,
since Rumford's time, that is to say more than forty years,
inasmuch as the fact upon which they mainly depended had
been known for that length of time.[1]

Mayer's peculiar cast of mind had remained uninjured
by his previous training; for this training had never included
a school or university course in physics or mathematics, so
that his original freedom of mind to follow his own line of
thought had never been interfered with, and books had not

---

[1] The electrical phenomena, which had been the most recent discoveries, did not give any further impulse in this direction; on the contrary, when not considered very thoroughly, they appeared rather to
suggest again the possibility of a perpetual motion, as in the form of an
electromagnetic motor driven by Voltaic elements. Even in the year
1841, the Swiss government had offered a large prize for an electric locomotive which would be cheap to run (it was even thought that the chemical transformations in the cells might provide valuable by-products).
This offer was withdrawn in 1844. See *Encyclopaedia Britannica*, Art.
'Perpetual Motion ' 14th Ed., vol. 17, p. 540.

been used for examination purposes, but simply as sources of facts. But a mental constitution of this kind is also the chief cause of the absence of all recognition, and hence of the unhappy fate of those who possess it. He who is not like the great majority of people, and hence does not appear adapted to the traditional form of life in the way in which he expresses himself, will not easily be understood, but will be readily passed over or even regarded with suspicion.

Mayer's first paper in Liebig's *Annalen* (1842) certainly received sufficient circulation; but it can hardly be doubted that the great majority of readers, and particularly of those who were not closely connected with research, could not have recognised the importance of its contents, if only on account of its shortness as compared with its richness in ideas. But it can just as little be doubted that all those whose minds were already busy with the question, which had been before science for forty years, must have been placed, immediately on reading the paper, or at any rate after short consideration, in possession of practically complete insight into the essentials of the new line of thought. It depended upon the few who proceeded to work on the subject and to spread the ideas whether Mayer would at once be recognised as the originator; and after the appearance of the second publication (1845), which was fully detailed, it was self-evident that no one had produced anything of equal value prior to Mayer. These few people failed in their duty.[1] We

[1] Helmholtz as Referee of the Physical Society in Berlin would have been best situated to report in the *Fortschritten der Physik* details of Robert Mayer's earlier writings as well as his own publication of 1847. But he only refers to Mayer so shortly, that one would suppose him to have published nothing of importance. Only later (1852 and subsequently), did Helmholtz gradually admit in public a few facts which had then already begun to be known to wider circles. (See in this connection Weyrauch's *Die Mechanik der Wärme*, Stuttgart, 1893, pp. 226-228 and 316; also the same author's *Robert Mayer*, Stuttgart, 1915, p. 67 ff., and Ostwald's *Grosse Männer*, Leipzig, 1909, pp. 272-274.)

Joule's first publication on the subject dates from 1843. William Thomson (Lord Kelvin) also put him forward as the originator of the idea of the equivalence between heat and work, in opposition to Robert

have only one exception to mention: Tyndall, Faraday's successor at the Royal Institution in London; he above all deserves the credit for the fact that Mayer finally enjoyed a few years of general recognition.[1]

Julius Robert Mayer was the third son of an apothecary at the sign 'Zur Rose' in Heilbronn; he early showed signs of a lively mind and receptive sense, and thus found for them plenty of food in his home, where there was much physical and chemical apparatus, a natural history collection, and books of all kinds. As a boy he spent a great deal of time in the open air, and also visited mills and factories in the

Mayer, but he expressly stated that he did so because Joule was a fellow countryman. This could not lead to much confusion, and we must also recollect that before Thomson made his report in England, Joule's work also remained unnoticed.

[1] For this reason we bring a portrait of Tyndall alongside that of Mayer. Tyndall was in many respects similar to Alexander von Humboldt; only as regards externals, a modestly reduced edition of him; Tyndall was a highly effective representative of, and propagandist for, the highest view of science and research. His lectures at the Royal Institution, which appeared for the most part in print, and also in the German translation, give a picture both charming and genuine of the state of scientific research at that time. In one of these lectures to a large audience drawn from all circles (Faraday was present) in the year 1862, he spoke in his own highly interesting style on energy and its transformation, and at the conclusion surprised his hearers with the remark, that everything of which he had spoken had been worked out quite independently by a German doctor, Robert Mayer, in Heilbronn, whose name was probably unknown to his hearers. He then added: 'When we consider the circumstances of Mayer's life, and the period at which he wrote, we cannot fail to be struck with astonishment at what he accomplished. Here was a man of genius working in seclusion, animated solely by a love of his subject, and arriving at the most important results in advance of those whose lives were entirely devoted to Natural Philosophy.'

Offence was taken in England over these remarks of Tyndall's and he was called to account publicly; but his opponents were silenced after his last reply (1864): 'To allow Doctor Mayer to remain in the position in which I found him would put the fault for that neglect upon me, from which only a reference to the ignorance of his contemporaries could liberate me. In every sentence which I have written in his favour, I was conscious of the force that only a completely original mind is able to offer, and without fear for his and for my faith, I now leave his reputation, and also my own behaviour with regard to it.' (See also Weyrauch, op. cit., pp. 338–342, also Kleinere Schriften, from which we learn that it was Clausius who made Tyndall acquainted with the too little known writings of Mayer.)

neighbourhood, showing a great deal of gift for understanding their mechanism. The question as to the possibility of perpetual motion, which no doubt at that time was generally discussed, early gave him much cause for thought when he heard of it at home. At school he was not regarded as a good scholar; but the old classics, and also Goethe's *Faust* remained favourites all his life. After leaving school, Mayer studied medicine at Tübingen, and then completed his training by visiting the hospitals in Munich, Vienna, and Paris; he did not find in his student days satisfactory lectures on physics at the university.

He began his independent career as a doctor on a Dutch East Indian ship with a crew of only twenty-eight, whose health was so good that he had little to do. There were also no passengers, so that Mayer was left practically alone with his thoughts for the whole journey, lasting from February 1840 to February 1841, and interrupted only by a stop in Java. But he had taken plenty of books with him, and, as he himself writes, 'he enjoyed a harmless peace of mind, which disposed him by preference to scientific occupation, and allowed him to lead a pleasant life, though in narrow circumstances and far from any companions of like taste; no day went by without interest of some kind.' In fact, everything made a great impression on Mayer, everything that he observed in the sky, on the water, or on the ship, as we see from his diary; and physics also, among the sciences, interested him from among his books, only he was not satisfied by the way in which the 'red thread' was interrupted in a thousand places, so that effects without causes, and causes without effects were presented to the reader. In particular, the inexplicability of the heat produced by friction attracted his attention, and also the production of heat in the living organism.

In Java he was obliged on account of illness among the crew, to let blood, and he was surprised and astonished to see that the blood was bright red, although it was not taken from an artery. This was the great event in Mayer's life; from this point everything that he had hitherto considered took on a new form in his thoughts; causes and effects appeared to him quite suddenly to enter into hitherto unknown connections. He then began for the first time – in the autumn of 1840 at Surabaya in Java – to think, in the manner to which we are accustomed to-day, about the energy principle. The whole hundred and twenty-one days of the voyage home obviously found him without anything else in his mind; for his diary, which on the journey out reports in detail all events on the ship and concerning the weather, is now completely silent, and a few months after his return he had completed a first draft on the subject.[1] But he continued uninterruptedly to work on the matter, pressing forward with ever increasing clarity, up to the statement of 1842, which appeared in Liebig's *Annalen* as the first public announcement of his thoughts.[2] We finally have the complete statement of 1845, which represents the highest point of his achievement in scientific research.

In the same year, after returning from the journey, Mayer took a house in his native town. He soon became the best-known doctor there, and had already married in 1842 the daughter of a well-to-do merchant. The following and no doubt happiest period of his life did not last long. We have already told above of the unhappy experiences which then overtook him. According to all trustworthy accounts[3] it was most of all the great mental isolation which Mayer

[1] He sent it to Poggendorf for the *Annalen der Physik*, where it remained pigeon-holed.
[2] See his letters of this period (Weyrauch, *Kleinere Schriften und Briefe von J. R. Mayer*, Stuttgart, 1893).
[3] See the collection of contemporary reports, together with Robert Mayer's own notes, in Weyrauch's edition of the *Mechanik der Wärme* (1893), pp. 303–309, and also the report on Mayer's meeting – 25 years after his treatment in the mad-house – with E. Dühring, in the latter's book *Robert Mayer, der Galilei des 19. Jahrhunderts* (2nd Ed., Leipzig), pp. 132–171.

experienced at home that resulted in a deep depression in view of the complete denial of all recognition, not of the truth of his ideas, but of their originality.

He hoped to find some consolation, together with suitable bodily care, in a sanatorium.[1] The first that he visited did not suit him; he was recommended another, where the young physician would have more time to pay attention to him. The advice was bad; the young doctor had just fitted himself out for a new treatment by means of a strait-chair, and applied this thoroughly to Mayer. An attempt to leave the sanatorium failed on account of periods of unconsciousness, which overcame Mayer in his state of complete weakness; he was brought back to the sanatorium and further ill-treated both bodily and mentally. The head physician, whom he wished to see, was never available; a Councillor, to whom he was taken approved the treatment, and his freedom was only obtained after thirteen months.[2]

His kind and trusting nature, which had been so grievously disappointed by both his scientific and medical colleagues, was completely transformed by his experiences; his iron will came uppermost; he felt capable of defying fate.[3] After a convalescence in Switzerland he was able to take up again his work as a doctor in Heilbronn. He always remained highly sensitive towards injurious and offensive remarks, and it must be remembered that plenty of opportunity of this kind still continued to occur; for a false report concerning him was spread in newspapers, public lectures, and finally (1863) even

[1] This double hope, which Mayer quite naturally entertained, is a good proof that he himself was an unusually good doctor for his own patient.

[2] From the statements of this Councillor, it appears that Mayer was treated for megalomania, for it was assumed that he imagined himself to have made a discovery which he had not really made.

[3] It should be particularly noted that Mayer continued to recognise willingly and joyfully all the achievements of his contemporaries such as Joule and Helmholtz, and even went too far in his respect on many occasions. Only, as he says, he was not prepared to admit 'that he had expressed any willingness to abandon documented rights of priority.'

in a well-known learned dictionary, which best shows the complete indifference shown in Germany towards him while still living. The report could not be rooted out; Mayer had died in a lunatic asylum ! The fact that Tyndall's championship of Mayer in London in 1862 finally broke the spell, a final consideration of which may be left to the calmer judgment of later times, has already been mentioned. Age brought Mayer still a few quiet and comfortable years; three well-grown children survived him; he died at the age of sixty-four.[1]

'As we stood around his grave,' says one of his biographers, 'a bitter feeling came over us. Suabia had given the world two men of science of the first rank; Johannes Kepler died as the result of want, while attempting to gain recognition of his rights by the Reichstag in Regensburg. Robert Mayer was misunderstood and insulted, until he was broken bodily and mentally. A foreigner brought about a change in his treatment at last. But he was now at rest; he had reached his goal.'

James Prescott Joule was an enthusiastic experimenter, and a wide-ranging thinker with the highest point of view. His most important work relates exclusively to the question of the nature of heat, which at the time when he began his work was still at the point reached by Rumford. In particular, the fact already recognised by Davy as of fundamental importance, the production of heat by the electric current, attracted his attention from the start. He investigated, and was the first to state the law of the production of heat by electric current, known under his name to-day, according to which the quantity of heat appearing in unit time in a conductor is proportional to the resistance of the conductor and the square of the current strength.

[1] E. Dühring, who had been banished from the University of Berlin, later (1877) did his best by means of public lectures in Berlin to enlighten his contemporaries. See his *Sache, Leben, und Feinde*, Leipzig.

His experiments in this case were of a simple description. He measured the current by means of a very simple galvanometer which he made himself, and which he graduated by means of a voltameter; the resistances were measured by introducing a copper resistance taken as a unit, and measuring the current, and then calculating by Ohm's law. The quantity of heat was measured by means of a simple water calorimeter; the most refined element of the whole equipment was a very sensitive mercury thermometer. In the case of metallic conductors he mentions, in connection with the dependence of the quantity of heat upon resistance, the discovery previously made by himself that discharges from the electrical machine produce an amount of heat roughly proportional to the resistance of the circuit; but he also carried out some measurements himself. In the case of liquid conductors this was done with greater elaboration, and both the development of heat in the elements, as also in electrolytic cells, was investigated at various measured currents and resistances, and found to be in good agreement with the law.

His measurements were not carried out with great accuracy. However, his experiments were clear in character; all subsidiary matters were recognised and taken into account; thus the development of heat taking place in the cells on open circuit was also deducted, and the polarisation voltage in the electrolytic cells, already discovered by Faraday, was correctly taken into account in measuring their resistance. It is always the case in the first investigation of natural processes, that great accuracy is not of so much importance as clarity in the experiments, and a basis of clear conceptions. Accuracy, and a consequent answer to the question whether further hidden phenomena may still be present, comes later, and often best indirectly.

Thus our present undoubting trust in Joule's law depends, apart from later and more refined direct measurements, mainly upon its agreement with the energy principle, recognised somewhat later and always found to hold strictly; this agreement could be demonstrated by the aid of Ohm's law. The three laws thus brought into connection with one another mutually support one another in the best possible manner, so that every new confirmation of one also reacts upon the other two. Nature is a connected whole; every new part that is rightly grasped always fits perfectly into what is already known, and thus actually simplifies our general view of the whole of knowledge.[1]

Joule had carried out these investigations on the heating effects of currents, with the special intention of seeing whether heat was thus actually generated, as in Rumford's friction experiment, or whether heat might possibly pass from the sources of current, the Voltaic cells, into the path of the current, in which case the elements should cool down. The result contradicted the latter view; the cells also became warm, in exact accordance with the resistance and the current. Joule already concluded from Faraday's electrolytic laws, that the total heat produced by the current in a closed circuit fed by Voltaic cells of any kind is proportional to the number of atoms chemically changed in the cells. He thus arrived at setting the heat of combustion, for example of zinc, in parallel with the current heat, which is connected in this way with the oxidation of zinc taking place in the cells.

It was therefore for Joule an obvious step, and one which followed directly from his line of thought, to proceed to find out whether the currents generated by induction according to Faraday's method produced cold anywhere in the

[1] It appears to be an increasingly popular device, unconscious or conscious, but in any case objectionable, to talk of 'classical' and 'modern' science, with the result that by introducing a division or opposition between two parts of our knowledge by means of such nicknames, the newer receives an apparent recommendation, instead of our admitting a want of clarity to be the simple cause of the appearance of a division or opposition.

circuit. He constructed for this purpose an induction machine with steel magnets, the rotating, current-producing armature of which was built into a calorimeter, while the current was led away to the outside to be measured. He found by comparison of the results on open and closed circuits, that here also only heat appears in the source of current, as in every other conductor, and he immediately regarded the heat appearing in the whole circuit as the equivalent, not of chemical transformation, as in the cells, but of the mechanical work used by the machine, just as in the case of Rumford's friction experiment. He now allowed the machine to be driven by falling weights, in order to measure the mechanical work, and he thus determined the number of foot-pounds which are equivalent to a calorie.

While Joule's law for the heat of the current had already been published before Robert Mayer's first paper (1840 and 1841), the experiments last described were published after Mayer, in 1843. The agreement, within the limits of accuracy then obtainable, of the equivalent found by Joule in so curious a manner, that is to say by introducing electrical current generation (calculated in metre-kilograms it was 460 for one degree centigrade and 1 kilogram calorie), with the result found by Robert Mayer in a totally different connection, namely by the warmth produced in gases (365 in the same units), must have immediately led to full recognition of the importance of Robert Mayer's discovery, and also of Joule's similar line of thought, that is of the energy principle, by all those who could understand it; there could be no further doubt that new insight of the widest possible range had been actually won into the interconnection of natural processes, as Robert Mayer believed from the start, and all that remained was to see how far the new ideas agreed with what was already known. This agreement was demonstrated in principle and on general lines by Robert Mayer in his

Us

second paper in 1845; it was Joule's endeavour to test the agreement by means of measurement in the most varied manner possible, and this occupied him for thirty-five years with experiments of continually increasing perfection and refinement, lasting from 1843 to 1878.[1]

It was a curious fact in this connection that Joule did not at first (1843) regard it as particularly necessary to carry out more elaborate experiments after what had already been discovered. He says rather – after he had also forced water in a calorimeter through narrow tubes, with measurement of the mechanical work required, and again obtained a fair agreement of the heat produced (423 mkg/cal) – that he did not wish to waste further time on experiments, in the conviction 'that the grand agents of nature are indestructible by the Creator's fiat.'[2]

It was only gradually, and also as the result of requests from outside, that he became more and more absorbed in the investigation, with great exactitude, of further processes of transformation. Thus in the period mentioned he measured the work done and the heat produced by compression of air, by friction of a paddle in water, by friction in oil, and in mercury, and finally once more by electrical heating, the latter with greatly refined apparatus. All the values for the equivalent agree with one another and with Mayer's value as far as the accuracy of measurement of the time allows us to expect, and the result of the two most careful investigations (425 mkg/cal) could for long be regarded as the best value known for the mechanical equivalent of heat, whereby Joule, originally a man of practical affairs with an enthusiasm for the highest type of scientific investigation, came to be

[1] From later and more exact measurements of the specific heats of air the figure 424 mkg/cal was found by Robert Mayer's reasoning, while later and more refined measurements by Joule of current heat gave 425 mkg/cal. The closer agreement in the two figures corresponds to the improvement in the methods of measurement.

[2] Philosophical Magazine series 3, Vol. 23, 1843; Collected Papers of J. T. Toule, Vol. i., pp. 157–158.

regarded as setting an example of the highest accuracy of measurement.[1]

James Prescott Joule was born at Salford near Manchester, being the second son of a brewer. The brewery was a family business, and had been founded by his grandfather. James Prescott was a weakly child, and was brought up entirely at home until his fifteenth year, when he began to work in the brewery, but at the same time received, together with his older brother, instruction in mathematics and science from Dalton. Immediately afterwards he also began to experiment himself, mainly with home-made apparatus, for which purpose he found room in his father's house. At the age of nineteen he wrote his first published paper, concerning a new kind of electric motor, which however he soon himself recognised to be of no advantage; at twenty-two his first publication on the production of heat by electric current followed. His father later built for him a laboratory. After he and his brother had taken over the brewery, he married, but his wife died in 1854, in which year the brewery, so far under his direction, was sold. From that time onwards he lived in great retirement, entirely devoted to his scientific work. From the year 1872 onwards his health was poor; from 1878 he received a pension from the Queen. He died at the age of seventy-one, at his birthplace, Salford.

Joule, like Faraday, had the rare good fortune to find in his youth a teacher who was worthy of him; in Faraday's case it was Davy, in Joule's, Dalton. Neither of them went through a special course of training 'in their science,' any more than almost all other great men of science; they produced their own science. In Joule's case we get the

[1] Later and still more refined measurements (which gave the value 427 mkg /cal) could only be carried out when the fundamental unit of heat, the calorie, had been more exactly defined, but this of course is only important in cases where the accuracy reached is of a most unusual description.

impression, as a result of his occasional remarks, that he read little of importance from recent times, and perhaps only kept himself informed by reports from technical journals. He also laid no claim to be regarded as an example of a learned man, but was, like Robert Mayer, the rare man of practical occupation, who sacrificed leisure and the richer enjoyment of life, or perhaps found the latter, in striving to come closer to the secrets of nature.

Hermann Helmholtz was the eldest son of a school teacher in Potsdam; he attended the school and gymnasium there, and then studied medicine in the Friedrich Wilhelm Institute in Berlin.[1] The means of his parents did not seem to allow of his studying pure science, whereas in the institute in question future military doctors were trained for low fees after passing as entrance examination. From 1843 to 1848 he was an army doctor in Potsdam, in which time he also passed his state examination in medicine. At the same time, he joined the new Berlin Physical Society, and soon became an abstractor to it. The manifold studies in literature, which he carried on at that time, also included fundamental mathematical work. He was also occupied with independent physiological investigations. At that time the physiologists and chemists were particularly interested in the question of the existence or non-existence of 'life force' and this formed the subject of Helmholtz's studies.

It was this that led to his publication in the year 1847 of a paper on the conservation of force, in which he carried the energy principle already founded by Robert Mayer and Joule, through all domains of physics, in parts going further than Mayer, and also stating it in mathematical form. In particular, we there have as a new discovery the fact that Faraday's law of induction likewise agrees with the

---

[1] For biographical details of Helmholtz see Tyndall's introduction to Helmholtz's *Popular Lectures on Scientific Subjects*, London, 1893.

conservation of energy, inasmuch as the cutting of lines of force, which is necessary to induction, always involves the expenditure of work, which is proportional to the induced current, from which it is seen that the heat of the current is the equivalent of the work done by induction (and not derived from the production of cold in some other place); Joule at the time had also shown that this was the fact. Helmholtz also shows in this paper that the law of conservation of energy may be deduced as a consequence of the exclusive existence of attractive and repulsive forces, the strength of which depends upon the distance apart of the bodies which act on one another – a proof, however, which has since lost the importance at that time ascribed to it, since it has become more and more clear that quite other kinds of force occur in nature, as Faraday already was aware.

In 1848, Helmholtz received a position as a teacher of anatomy in the Berlin Art Academy, and a year later a professorship of physiology at the University of Königsberg, whereupon he married. He there invented the ophthalmoscope, which enables the retina of the living eye to be seen. In the year 1855 he went as professor of anatomy and physiology to Bonn, and in 1858 as professor of physiology to Heidelberg, where he married for the second time after the early death of his first wife. In the year 1871, he was called to Berlin as the first professor of physics, and in 1888, at the age of sixty-seven, he became president of the newly founded Physikalisch-Technische Reichsanstalt,[1] which post he retained until his death at the age of seventy-three.

Helmholtz's scientific individuality lay in his wide ranging capability of easily making himself at home in all domains of knowledge; the essentials were soon grasped by him, and were then ready to hand for further work. All the great connections within exact knowledge of all kinds had to be familiar to him in order to render possible his numerous and

[1] Corresponding to the English National Physical Laboratory.

manifold researches, which are mostly of a kind developing existing knowledge of nature and connecting it together and enlarging it.[1]

The wide range of his gifts included a rare aptitude for mathematics; he thus succeeded in deriving, from the differential equations of hydrodynamics, the characteristic phenomena of eddy formation, and of the structure of jets in liquids and gases, in the most admirable fashion. This was not the achievement of further knowledge of nature, but an unusual mathematical success.[2] For the fundamental equation had already existed for almost a hundred years, and eddies in liquids and their flow as jets had been known for a much longer time; but no one till then had succeeded in showing that these phenomena are not only contained in the fundamental equations – and therefore do not conceal anything fundamentally new, but follow entirely from Newton's and Galileo's laws of motion, as Newton had already seen – but also that they may be completely deduced in detail from the fundamental equation, and represented as they occur, according of course to the fundamental laws, but in gases and liquids having particular properties. This at the same time made clear, in a manner superior to all observation and going beyond Newton's earliest statements, what is the essential and characteristic feature of these phenomena, what are their simplest forms, and what can be further done with them, or thought about them.

The fact that this achievement was reserved for Helmholtz, who had never studied mathematics at the university at all, shows, in a striking manner, the complete uselessness of the extensive mathematical and other courses of

[1] See his collected papers (*Wissenschaftliche Abhandlungen*) Leipzig, 1895. English translations: *On the Sensations of Tone*, trans. A. J. Ellis, London, 1875; *The Description of an Ophthalmoscope*, trans. T. H. Shastid, Chicago, 1916.
[2] See what has already been said in this connection under Laplace and Gauss.

training at present-day universities, in which innumerable
students are plagued with the most out-of-the-way matters
merely for the purpose of examination, and many of these
again, as teachers in schools, pass on to the next generation
the same endless and useless plague, instead of inculcating
modest but accurate fundamental knowledge and the simple
great fundamental sense of mathematical thought, whereas
only few are capable of originating any kind of progress by
means of mathematics, and have no need to waste their time
in this way.

As an example of Helmholtz's gift of correctly deducing
the possibility of unknown natural phenomena from known,
we may mention his early reference to electrical oscillation,[1]
six years before this process, so well known to-day, had
been calculated by William Thomson on the basis given by
Faraday, and ten years before it was actually observed.
Then we have his proof that light of sufficiently short wave-
length would pass through everything in straight lines, three
years before rays for which this is true were discovered, and
twenty years before the rays so discovered (and quickly
used in medicine), were actually shown to be ether waves of
extremely short wave-length. Helmholtz then also produced works of a general description, which contributed
greatly to his public reputation, some of these being of a
popular character. We may mention his larger works, the
Physiological Optics, and his *Theory of Sound*, and also many
lectures and papers, in which his understanding for art, and
particularly for music, played a part, and in which he was
able to gain a very large circle of readers, since he met the
taste and views of his time halfway, as for example, when he
allows himself to give an estimate of Aristotle.

[1] In the paper on the conservation of force, 1847.

RUDOLF CLAUSIUS (*1822–1878*)
WILLIAM THOMSON, LORD KELVIN (*1824–1907*)

AFTER the principle of the conservation of energy had been recognised, and its limitless validity had become increasingly more evident, it became time to follow out further Carnot's ideas concerning the effect of heat, in the direction in which he himself was already pursuing them, as we have learned much later from his posthumous papers. The prosecution of this plan was for the most part the work of Clausius in Germany and W. Thomson in England, though both of them were the authors of other by no means unimportant achievements. Clausius, who was the older, was also the pioneer, and Thomson immediately followed, and supplied a great deal of additional matter, which extended to the most out-of-the-way conclusions. It was the recognition of a new fundamental law relating to heat which was here the essential step forward, cleared up Carnot's doubts, connected together into a single system older and more recent knowledge of heat, and allowed many further conclusions to be drawn.

This new fundamental law introduced by Clausius is called the second law of the mechanical theory of heat, or thermo-dynamics, the first law being the energy principle. The second law is also founded upon experience; but it was not necessary, as William Thomson at first thought, to collect further experience in order to arrive at a new point of view. It is only necessary, as Clausius could show, to make use of a law derived from existing experience, namely: 'that heat never passes of itself from a colder to a warmer body.' The fact that this statement is correct is shown by all transference ofh eat, by means of conduction, convection, or radiation, not however without it being necessary to give to the words 'of itself' a particular suitable interpretation. This law of experience can already be described as the second law of the theory of heat; but it may also be given other forms, in particular such as are directly suited to calculation; this was done by Clausius.

For this purpose he introduced a new quantity, the 'entropy,' which he defined, but which we cannot here discuss any further. We can only remark that the second law also enables us to determine what fraction of the heat supplied can be transformed into work in a heat engine of the best possible description, that is to say, as Carnot showed, one completely reversible. This fraction is given by the ratio of the available range of temperature to the absolute temperature, that is to say, the temperature of the body reckoned from $-273°$ C, at which the heat to be transformed is taken. This result was of great importance for all progress in the construction of heat engines, from the steam engine to the most recent type of internal combustion engine.

Regarding other results of thermo-dynamics, we can only mention the deduction of the relationship between melting point and pressure, and the discovery of the peculiar properties of saturated vapours. The former is of importance in connection with our estimate of the state of the earth's interior, and also our knowledge of glaciers, the latter regarding the finer observation of the processes connected with water vapour in the earth's atmosphere, and in the steam engine.

Before we go into further achievements of these two men of science, we will shortly consider their lives.

Clausius was born in Koslin in Pomerania, being the sixth son among the eighteen children of a school inspector. He attended the gymnasium in Stettin, studied in Berlin, went early into teaching as a means of assisting the education of his younger brothers and sisters, became a member of the university of Berlin at twenty-eight, and five years later was called to the technical university in Zurich, where he remained for twelve years and founded a family. He then became professor at Würzburg, and soon afterwards at

Bonn, where he remained for the rest of his life. When at Bonn he also took part in the campaign of 1870; a wound in the knee occasionally caused him trouble, but he was still able at the age of fifty-six to ride as a means of exercise. He died at the age of fifty-six. Just as in the case of Robert Mayer, contemporaries also report of Clausius, even in his schooldays, his strict love of truth, trustworthiness, straightforwardness, and devotion to duty, as particularly characteristic qualities.

William Thomson, though Scottish in origin, was born at Belfast, in Ireland. His father was a teacher of mathematics there; he undertook the entire education of his son up to his tenth year, and in a great many directions. He was then called to the university of Glasgow, where he entered his son at this early age among the students. The result was satisfactory; William paid particular attention to Fourier's work on the conduction of heat, which is full of points of view new at the time, and of mathematical art, and also to Laplace; he soon also published a first small paper in connection with Fourier. At the age of sixteen he then went to Cambridge and later for a year to Paris, where he worked in Regnault's laboratory.[1] Shortly after his return to Glasgow, the professorship of natural philosophy became vacant, and William Thomson, at that time only twenty-two years of age, received the post. With very modest means, he immediately laid out an experimental laboratory in an empty wine cellar; it was not until 1870 that a new laboratory was built.

Thomson remained faithful to Glasgow for the rest of his life; together with a pleasant and stimulating degree of teaching activity, he had leisure for scientific work; he was able to overcome the smallness of the means at his disposal by exploiting his technical inventions. Among these were

[1] Regnault, born in 1810 at Aix la Chapelle, died in 1878; was famous for his exact calorimetric measurements.

his quadrant electrometer, which soon became indispensable, together with various forms of direct reading ammeters, which at that time were a complete novelty, and his exact appliances for investigating atmospheric electricity; furthermore, his improvements in the ship's compass and apparatus for taking soundings. But one of his most important inventions was his receiving apparatus for cable telegraphy (1867), the well-known siphon recorder, which for a longtime was used exclusively for reception from long distances; such reception was only made possible by its use.[1] These achievements, so rapidly made and of such general interest, no doubt gave much satisfaction to Thomson himself, though they could only be for him a subsidiary matter; but they had without doubt the greatest influence upon his public reputation and also upon his elevation to the peerage, which occurred in 1892, when he took the title of Lord Kelvin.   They also enabled him to maintain a large country house and a yacht.   A wife who always took the greatest care of his well-being was his lifelong companion.   At seventy-five he gave up teaching; he died at the age of eighty-four.

A particularly important achievement of Thomson's was his first calculation of electric oscillations.   He was the first to make clear how the self-induction of a circuit discovered by Faraday must result in the production of alternating electric currents in an open circuit of conductors, and he had calculated in every detail, from the law of induction and Ohm's law, the nature of such oscillations; that is to say their period, damping, and the whole course of their intensity, as dependent upon the capacity, self-induction and resistance.[2]

Such oscillations, the existence of which had already been suspected by Helmholtz, could then actually be produced

[1] The laying of the cable, which was also not a simple matter was first successfully accomplished by Werner Siemens; see his memoirs.
[2] *Philosophical Magazine*, Vol. 5, p. 393, 1853.

and examined. Maxwell added the idea of electric waves, which should emanate from such oscillations, and showed what the properties of these waves should be, and Hertz, somewhat later, was actually able to discover and study these waves. All these discoveries we shall still have to consider in their chronological order.

In order to form a true estimate of Clausius and Thomson, but particularly of the former, we must now once more turn to the phenomena of heat.

Clausius was not only (from 1850 onwards) the founder of thermodynamics, but also (from 1857 on) developed the kinetic theory of gases. While the former starts from the two laws already mentioned, according to which heat is treated as a form of energy without further enquiry into its nature, the latter discusses the motion of the molecules, which is the essential nature of heat; here heat actually appears as that which Rumford already supposed it to be, namely motion. Knowledge of nature was now sufficiently advanced for this to be asserted with certainty, and also for the nature of the molecular motion to be stated. This could be done most completely for the gaseous state of matter. Joule had already (1851) made a well-founded beginning of this, which was no longer a simple and arbitrary hypothesis, but was based upon his studies on the mechanical equivalent of heat.[1]

Clausius now calculated, in a manner free from all objection, the velocities of the molecules and later their free paths, that is to say, the distance through which they move in a straight line between collisions. The velocities are very great, the free paths very small at ordinary temperature and pressure, so that the collisions of the molecules are very frequent. These collisions take place according to the laws already founded by Huygens for perfectly elastic bodies.

[1] Joule's *Collected Papers*, Vol. 1, p: 290: 'Some remarks on heat and the constitution of elastic fluids.'

The temperature measured from absolute zero $(-273°\,C)$ is shown to be proportional to the kinetic energy of the gas molecules; the pressure of a gas is the simple effect of the collisions of the gas molecules with the walls of the vessel. The tendency of every gas to spread through all space, already recognised by Guericke, is simply a consequence of the straight line motion of the molecules according to Galileo's inertia law, and is not caused by repulsive forces between the molecules. The gas laws of Boyle and Mariotte, of Gay-Lussac and Dalton, and of Avogadro, then follow of themselves as purely mechanical consequences of the molecular motion. Avogadro's law was first put beyond all doubt as a result of these connections, by its being linked with a large number of well-established facts, both new and old.

All this also led to Dalton's and Avogadro's views on the atomic structure of matter, and the union of atoms to form molecules, almost suddenly acquiring practically complete certainty, which up to that time had been gradually approached throughout fifty years; and upon this basis a very great deal more, now almost limitless in extent, was built up.

Furthermore, it was not only possible to state the relative weights of atoms and molecules, towards which science had been feeling its way since Dalton, but to estimate their diameters in ordinary units of length, and soon also their absolute weights, and hence the number of them contained in a cubic centimetre of every gas at given temperature and pressure. The internal friction (viscosity) of gases, their diffusion, and their heat conducting power, already studied by Newton and Coulomb, could now be understood in every detail and also quantitatively, with the result that their peculiarities, in part quite unexpected in nature, could be predicted, and then confirmed by measurement. For example gases, as opposed to liquids, increase in viscosity when heated, that is to say, become less fluid. The result was that quantitative relations were obtained, which could then be used in a converse sense for still further investigation of the peculiarities of the gas molecule. The specific heats of gases, with their peculiar laws, could also now be understood, and thus became a new means for the accurate estimation of the number of atoms in a molecule. It was found possible to distinguish monatomic molecules, such as those of the metallic vapours (and later the rare gases), from di- and poly-atomic molecules. All this was put upon a firm basis by Clausius. The further rapid development of these matters, which was completed as regards essentials within ten years, was due, besides to W. Thomson, to Maxwell, Boltzmann, and Loschmidt.[1]

After this development in our knowledge, an old problem was also finally solved, namely that of the liquefaction of gases. The present method of manufacturing liquid air on a large scale depends upon the discovery made by Thomson in collaboration with Joule in an experimental investigation carried out with great persistence, and we will now discuss this. Originally, all the gaseous substances discovered by Scheele, Priestley, and Cavendish gave rise to the fundamental question whether they could by any means be reduced to the liquid condition.

Faraday was the first to work on this problem, which he did in 1823, when still Davy's assistant. He made use of the means which had been known since Dalton's investigations of vapours, for the liquefaction of the latter: pressure and cold. He succeeded in liquefying by simple means chlorine, carbonic acid and other gases. He caused the gases to be generated in sealed glass tubes, and thus produce their own pressure, and cooled one end of the tube, which was bent downwards, in a freezing mixture. It was then found out that this liquefaction could be carried out on a large scale by

---

[1] Loschmidt was born in 1821, and was the son of poor peasants near Carlsbad in Bohemia; he died in 1895 as professor of physics at the University of Vienna.

means of pumps; but nevertheless, this process failed, even with the highest obtainable pressures, in the case of a number of gases, such as oxygen, nitrogen, and hydrogen.

Faraday therefore took up the investigation again in 1844; now starting from observations published since the time of his first experiment by Cagniard de la Tour, a French engineer.[1] In his experiments, liquids such as ethyl ether, carbon bisulphide, and water, were heated in closed glass tubes, and it was observed that they are completely transformed into vapour above a certain temperature, although the pressure, which was measured accurately in several cases, becomes very high, and although the density of the vapour as calculated from its volume cannot be much smaller than that of the liquid. He determined for each liquid a definite temperature, later called the critical temperature, above which they did not remain liquid in spite of the highest attainable increase of pressure; for example, this temperature for ethyl ether was 187° C.

Faraday realised that this must also be of importance as regards the converse problem, namely the liquefaction of gases, which would accordingly be more a question of sufficiently low temperature than of the degree of pressure used. He therefore made a large number of new experiments with many gases, in which he made use of simple pumps, and a freezing mixture of solid carbon dioxide and ether, which at that time was already known, and the effect of which he further increased by pumping off carbon dioxide, whereby a temperature of about —100° C was reached.[2]

He thus succeeded in liquefying a further number of gases; but oxygen, nitrogen, hydrogen, nitrous oxide, and carbon monoxide still resisted liquefaction. Faraday also investigated the vapour pressure of the various gases which he

[1] Cagniard de la Tour (1777–1859) was also the inventor of the siren which has become important for acoustical investigations, and is now also used for signalling on ships, and so on.

[2] Described in the *Philosophical Transactions of the Royal Society*, 1845.

liquefied, and its dependence upon the temperature; he concludes in the case of carbonic acid that its 'Cagniard de la Tour state' (critical point) lies in the neighbourhood of $90°$ F. ($32°$ C) at which the gas and the liquid are equal in density, and that it therefore does not become liquid at higher temperatures, whereupon he points out that the other gases which he could not liquefy must require the use of still lower temperatures than he was able to produce. This result was confirmed fully twenty-six years later (1873) by Thomas Andrews,[1] who followed with still greater exactitude and much improved apparatus, the connection between pressure and volume of carbon dioxide at different temperatures; he found, in surprisingly good agreement with Faraday, $31°$ C to be the critical temperature of carbonic acid, and gave detailed curves showing its complete behaviour.

Upon this basis, Van der Waals (1837–1923; lived in Leyden and Amsterdam) immediately put forward the admirable equation known under his name, which summarised the relationship above stated in a simple manner holding for all gases, and also giving the long known departures from the gas laws of Boyle and Mariotte and also of Gay-Lussac and Dalton; it also summarised all questions of the liquefaction of gases (1873). This achievement represented the interconnection of a large number of single facts, which could thereby also support one another. This equation further connects in the best possible way with the results of the kinetic theory of gases, which likewise leads to perceptible departures from the simple gas laws, as soon as the actual volume occupied by the molecules becomes appreciable in relation to the total volume of the gas, or when the forces begin to come into play, with which the gas molecules attract one another. Van der Waals' equation gives

[1] 1813–1885; began as a practising physician, and then became professor of chemistry at Belfast.

for every gas a special figure for this molecular volume and for the molecular forces. The molecular forces are, in agreement with Laplace's theory of capillarity, very small in the case of gases, since the gas molecules when in motion are at an average distance apart much greater than the range of action of these forces; but they do not entirely vanish.

The existence of small molecular forces in the case of gases could already be concluded, long before this time, from the remarkable investigation of Joule and Thomson in the years 1852–1862, which we have already mentioned. The investigation showed that cooling took place when a gas expanded without doing external work. According to the energy principle, such cooling could only take place when work was done in some way within the gas itself by the increase in volume, which work would then result in a corresponding disappearance of heat. This would be the case if attractive forces between the molecules existed. The experiments carried out in 1806 by Gay-Lussac had not shown any cooling; this cooling, and hence the effect of the molecular forces, could thus in any case only be very small. Joule and Thomson finally succeeded, after overcoming many difficulties, in actually demonstrating a small decrease in temperature (a few tenths of a degree), by pumping the gas in a circle, and arranging that it should expand suddenly at a constricted part of the pipe line. External work was not done by the gas, since its total volume in the circuit remained unaltered; nevertheless, the small fall of temperature already mentioned was found to take place at the constriction. The presence of molecular forces, but also their smallness in the case of gases where the molecules are far apart, was thus proved, in full accordance with the small departures from the laws of Boyle and Mariotte.

When so much information had now been obtained concerning gases, the ground was well prepared for an attempt to liquefy oxygen, nitrogen, and also atmospheric air; this

Ws

was successfully accomplished in 1877. The higher degree of cooling, already recognised by Faraday as necessary, was attained in these very excellently conducted experiments by very simple means, already known since Dalton's time, of allowing the strongly compressed gas, cooled as far as possible, to expand suddenly.[1]

Although liquefaction only existed for a short time in the form of a mist or a few drops, it was actually seen, and the certainty that liquid oxygen exists was obtained. It was now settled that the opinion generally held by the uninstructed public after the failures of so many years, that air is a 'permanent,' or 'incoercible' gas, was incorrect. After the existence of molecular forces, however small, had been proved, this possibility could have been regarded as a certainty; for molecules which have any perceptible attraction for one another, must, at a sufficiently low temperature, and hence at a sufficiently small distance apart, cohere to form a liquid, and finally a solid, as a result of the correspondingly increased molecular attraction.

The method just described, and also that used in ordinary refrigerating machines, was not suitable for the preparation of liquid air on a large scale; here the experiment of Joule and Thomson, carried out in a suitable manner, leads to success. This experiment, in which the gas is pumped round in a circle, only produces a very small fall in temperature, and it might therefore have been regarded as entirely unfitted for the purpose; but Joule and Thomson had shown that the small fall in temperature at the constriction increases considerably when the gas is cooled; this is readily understood, since at the lower temperature the molecules are closer together, and hence their attractive force is greater. Accordingly, in the present liquid air machines, the circuit of

[1] L. Cailletet in Paris; he had already carried out, before Andrews, measurements on gases at high pressure, which Van der Waals was able to make use of in support of his equation.

the gases is so arranged that the gas cooled after passing the constriction flows back over the pipes carrying the gas arriving at the constriction, and thus cools it. In this way, the fall of temperature continually increases if sufficient time is allowed, until the critical temperature ($-119°$ C in the case of oxygen, $-146°$ C for nitrogen) is reached and passed, and from this point the apparatus begins to deliver a continuous stream of liquid air, it being of course necessary to continually replace the liquid drawn off by a corresponding amount of gaseous air. One litre of liquid air requires about one cubic metre of gaseous air, which fact gives us a direct idea of the comparatively great molecular distance in the gaseous state as compared with the liquid state. The success of this method is one of many examples of the fact that every natural phenomenon, even though at first it appears in so minute a form as hardly to be perceptible, can nevertheless be raised in its effects to any degree, as soon as the laws of it are known.

After air, hydrogen and the most hardly liquefiable gas of all, helium, have been liquefied, and finally also solidified. Although the difficulty was very great, no new fundamental knowledge was gained by the process. However, the liquefaction of these gases gave us the means of carrying out entirely novel investigations at the lowest temperature; for if liquid helium, for instance, is being used, we are certain of maintaining steadily a temperature which cannot be greater than the critical temperature of this substance $-268°$ C, that is only $5°$ C above absolute zero, below which the temperature scale does not extend, since at that temperature the motion of the molecules ceases, and their kinetic energy, which is the measure of their temperature, becomes zero.

CHARLES DARWIN (*1809–1882*)
CARL LINNAEUS (*1707–1778*)
AND THE INVESTIGATION OF LIFE BEFORE AND AFTER HIS TIME

LIVING matter only became an object of science at a late period. The main reason for this is certainly a reluctance to disturb living things, to interfere with them, not to say to destroy them, when they do not confront us as enemies; a reluctance which is particularly characteristic of the type of humanity which is inclined to scientific investigation. The purely utilitarian object of increasing knowledge does not suffice the human mind of a high type, and cannot reconcile him with the idea of torturing an animal or even of making use of a corpse; what has, or has had life has always, and in all ages, been sacred to him. But the coming into play of a much higher intention, namely that of actually aiding life by means of knowledge thus obtained, finally induced men with high gifts to examine thoroughly what goes on in the interior of the animal and man.

It was thus the art of healing which led to a science of living things (anatomy, physiology, and biology), when it became more and more clear that life could not be effectively aided without such knowledge. Assuredly also, the quite obvious complexity of all processes taking place in living matter was a great hindrance, and this must have been sufficiently alarming, as long as even such a simple process as the fall of a stone had not been grasped in all its details. However, not only had all lifeless machines been already well investigated from Archimedes to Stevin, but even the laws of motion of the planets in their orbits had already been found by Kepler, before anything at all was known of the circulation of the blood in animal and man; indeed such a circulation was almost refused any consideration. It was a physician, William Harvey (1578–1657) in England, who first elucidated the movement of the blood, and the function

KARL LINNAEUS

CHARLES DARWIN

of the heart as a central pump (1628), long after water pumps with pipes had become quite familiar, while the ideas current concerning the heart and the veins had remained confused and unclear.[1] This shows how slow was the progress in the investigation of living things.

This example also tells us that people were for a long time disinclined to apply experience gained with non-living things to living. It has always been an unmistakable fact that the living creature has something peculiar to itself, which in human beings and the higher animals is so obvious in the form of freedom of will, and that this is wanting in the case of the non-living. The peculiar difference was no doubt accounted for by assuming the two worlds, the living and the non-living to be totally different from one another, and no one was willing to imagine that the same laws of nature could hold for both. In a similar way, there was a reluctance to assume the same laws of motion for celestial and terrestrial bodies, until Newton was able to prove beyond doubt that these laws really are the same. We have already shown, in treating of the age which followed Newton, that the revelation of this fact soon led to spurious 'enlightenment,' and accordingly, living things were now regarded, quite in contrast to earlier times, as pure mechanisms, of the same nature as non-living things; although it was not known whether really all non-living things were pure mechanism, such as were known from the investigation of material bodies.

At that time, the physics of the ether was only, as a result of Huygens' work, in its initial stages; nevertheless it was right, since in any case necessary, first to get to know from all sides the bodily machine of living things, and to investigate

---

[1] See the *Works of William Harvey*, London, 1847, particularly chapter 8, page 45. Harvey, the physician, also gave expression to the fact that every animal has its origin in some egg-like structure (*omne animal ex ovo*). The sexual reproduction of plants and the mechanism by which this is effected, was only satisfactorily determined about sixty years later (Camerarius in Tübingen, 1694).

it as thoroughly as possible. What goes beyond this could never be quite forgotten by the more farseeing, in particular by physicians, though it might be called, as already by Socrates and Plato, and then by Paracelsus, Descartes, and later writers, some such thing as soul, spirit, *archaeus*, or 'life-force,' and though recognition of it by the learned might pass through the greatest possible variations.[1] The investigation of the mechanism of living things made use of all the means and knowledge already provided by scientific research in general.

In the course of two centuries, progress was made, from Harvey to Wilhelm Weber, in the investigation of the coarser mechanisms; the latter together with his brother, investigated for the first time the blood wave which causes the pulse, and must be distinguished from the flow of the blood, and the mechanics of locomotion. In the case of the finer organisms, the application of the optical investigations of Galileo and Kepler ever-increasing improvements in lenses and microscopes lead to starting from Toricelli's simple glass ball, which was so long used with success; in this way smaller animals also could be investigated as regards their internal anatomy, and their manner of life and reproduction, and the smallest forms of life were discovered. Already in 1650, Leeuwenhoek and Swammerdam[2] proceeded to investigate the metamorphosis of insects, and to discover the blood corpuscles and the infusoriae; and a hundred to two hundred years later, the cellular nature of the structure of all living matter was made clear, the cells and their contents proving to be the essential carriers of life.[3]

[1] The English reader may be referred to W. McDougall's *Body and Mind: A History and a Defense of Animism*, London, 1911.

[2] Leeuwenhoek lived between 1632 and 1723 in Holland; he used only simple lenses which he ground himself. Swammerdam lived between 1637 and 1680, likewise in Holland.

[3] Here the very gradual progress made appears to have depended almost entirely upon the gradual progress in perfection of the appliances and of the use of the microscope.

In the finest elements of structure, right down to the atoms, progress was made in the investigation of living matter by the aid of the chemical knowledge obtained by Scheele, Dalton, Davy and Berzelius. New substances with characteristic properties were continually being discovered and investigated; they were taken from plants and animals, separated, and prepared in a pure state. Scheele himself had already made a good beginning, when he prepared in a pure form such substances as tartaric acid, citric acid, and malic acid, and found out how to distinguish them one from another. These substances were then analysed, and no other elements were found in them than those already known from non-living nature, the chief being always carbon, and along with it almost always only hydrogen, oxygen and sometimes nitrogen; otherwise surprisingly few further elements. In view of this uniformity, it became all the more important to discover the differences in their quantitative composition, for which purpose Berzelius, and twenty-five years later (1837) in a more exact manner, Liebig, developed suitable methods particularly adapted to the case of these few elements.

Berzelius was already able to show that in organic compounds, as the substances occurring in plants and animals were early called, fixed proportions by weight of the elementary components also hold good, and can be expressed in the numerical relationships of Dalton's atoms. The numbers of atoms which had to be regarded as combined with one another in these cases, were often fairly large, and hence difficult to determine with accuracy; but gradually the recognition of the recurrence of certain smaller groups of atoms (radicles) as a common constituent of the larger groups, was of great assistance. The structure of the latter could thus be determined without dependence upon quantitative analysis alone, and without the determination of molecular weight, for which purpose the methods given by

Gay-Lussac and Avagadro, as well as those discovered later, often failed in the case of these organic bodies.

This led to an ever-increasing number of substances containing ever-increasing numbers of atoms in the molecule, and of ever-increasing complexity becoming known, and this forms the substance of the sciences of organic and physiological chemistry. These substances, all of them compounds of carbon, were in part taken from living creatures, but also in part had gradually been made artificially from their elements without the intervention of life; they were thus as it were imitated from life, and finally many more substances were made, at least of a simple description, than are actually discoverable in plants and animals.[1]

However, the largest molecules and those containing most atoms, which still provide so many questions for chemistry, are always found in living substances; they appear to form that part of the cell most important for life.

If we thus glance over all that scientific research has discovered in detail concerning living things, little of a fundamentally new description appears. The only new fundamental fact concerning the nature of living matter is that the very largest molecules, containing an enormous number of atoms, form the seat of life with its wonderful phenomena, which obviously fall, as regards their origin, outside the frame of all our previous investigation of nature. Everything else which depends upon these centres of life, and proceeds from them, is more or less comprehensible on the basis of our experience with non-living matter. We recognise that these centres make use, in the most varied way, of all the processes which the physics of matter and of ether have already

[1] A particular example of this, which one often finds put forward as a milestone on the progress of chemistry towards living matter, with exclusion of a life force, is the synthesis of urea without the use of urine, which Wöhler (a pupil of Berzelius) succeeded in accomplishing in 1888 in Berlin. It must be remembered, however, that urea never appears as a carrier of life; it is a waste product of life, and is as dead as carbonic acid or water vapour and all other products of chemistry hitherto made.

revealed to us, in order to bring about, purely by guidance or release, everything that happens in a living organism. Much of this is quite coarse mechanism, not going beyond our knowledge of the physics of matter, for example the lever action of the skeleton, and the pumping action of the heart.

In finer structures, for example the conduction of nervous stimuli, electrical processes have been recognised as taking part; the working muscle is undoubtedly a molecular mechanism, acting through molecular and atomic forces which can also be studied in non-living matter, and which, at least in part, are of an electro-magnetic nature and so belong to the physics of ether. The peculiar fact is simply that the production and continual fresh formation of such well-arranged molecular mechanisms in the nerve and muscle, and indeed the synthesis of the necessary molecules from lifeless groups of atoms, such as their food presents to living beings, can only take place in the presence of the aforementioned centres of life. Nothing of the kind has ever been observed, nor has life itself therefore ever been found, without a previous living germ produced by a living organism.[1]

[1] We may here indicate a point of view which occurred to the author more than twenty years ago from the consideration of what we know of life, but which must also occur no doubt to everyone, who, with the whole content of the physics of matter and of ether in his mind, regards as the first task of research in breaking new ground, the formation of notions so adapted to the actual and observable process of nature, that the latter becomes as comprehensible as possible by their means. This view regards as quite possible the new formation of life, the spontaneous generation hitherto sought for in vain, as soon as sufficient suitable molecules containing enough atoms are available, and in an environment which is suitable and allows of metabolism. What is then necessary for such molecules, forming a body, in order for them to live, is a suitable spirit. Spirit is here a name for that which is obviously necessary, apart from the material, to life. A name that enables the concept thus grasped, the thing discovered – unknown but still existing – to be firmly retained, in order that it may be investigated as far as possible by further experiment. It follows from this that we must assume as existing in space in an available form spirits of many different kinds (derived from dead living beings), and that these have the property of uniting with molecules which suit them, as soon as they find such, whereby these become living beings or the germs of them, which then develop in a suitable environment in accordance with the spirit and guided by it. These are ideas which are

It is particularly noteworthy that the general law of the conservation of energy has been found to be followed by all living beings, a fact already thoroughly considered by Robert Mayer. He already pointed out that the muscle transforms the chemical energy of the food brought to it by the blood directly into mechanical energy, whereas the steam engine, and all similar motors, make use of the intermediary of heat. The discovery of the second law of thermo-dynamics, by Carnot, Clausius, and Kelvin, has made this a certainty; for in the muscle the differences of temperature are absent, which are essential for the action of heat engines, and which determine their efficiency. The efficiency of the molecular muscle machines is therefore actually much greater than that of the best heat engines.

Summarising what we have said so far; life is, at the last resort, bound up with peculiar molecules, which are contained in the cells, in structures visible under the microscope, and beyond this fundamental secret, the phenomena of life do not offer anything fundamentally new in kind. The enormous industry of observers has further brought to light an almost limitless number of single facts, relating to the various species of plants and animals, their constitution, and their life history. Here it is of the highest importance to order the facts in an accessible manner. The first great constructor of systems and summariser of knowledge in the domain of living things, and on the basis of his own observation, was Carl Linnaeus; but he only appeared fifty years after Newton's *Principia*.

Linnaeus was born in 1707 as the first son of a country clergyman near Verjo in the south of Sweden, and studied

suited to a comprehension of life, with its coming and going of ever new, and also differently constituted, forms, and with its phenomena, which so obviously transcend everything observed in lifeless matter and in ether. These ideas could be of benefit for the further investigation of life, in the first place chiefly of the sub-microscopic forms which must be assumed, without any injury to the continued pursuit of what is more directly accessible to the senses.

medicine, first in his birthplace and then in Holland, but he always paid particular attention to plants. He then practised in Stockholm, where he married, until in 1741 he received a professorial chair and the botanical garden at Upsala, where he remained for the rest of his life as regards the main part of his work. But repeatedly, also in his earliest youth, he undertook expeditions, and proved a tireless and wide-ranging observer of nature, and collector of plants, animals and minerals; which activity, combined with a wide study of literature, made him an incomparable expert regarding all kinds of living beings. In later years, when he had already become famous, he was then further aided by receiving all kinds of natural objects from all parts of the world.

Gifted with a rare power of taking a wide view, he was able to create, in several large works, a classification and nomenclature for all living beings, which was immediately generally accepted, and acted upon botany and zoology like a new creation of their whole content. Everything that lived now had a fixed name, and was fitted into classes, orders, families, species, and varieties; a complete survey was now possible of the mass of single facts which had become quite unmanageable; the sciences of life had also likewise received a useful artificial language, which was actually an essential preliminary for all further prosperity, since it enabled one botanist or zoologist to understand others. Hence the recognition which Linnaeus, ennobled under the name of Linné, already received during his lifetime was very great. The Queen of Sweden, the sister of Frederick the Great, and also her son (Gustave III), showed him the greatest respect and assistance. He died in 1778, aged seventy-one.

It would be wrong to regard Linnaeus merely as the great systematist, as was done for a very long time.[1] Although his works are by far the greater part, and in the first instance

---

[1] I must here acknowledge my indebtedness to Professor E. Almquist in Stockholm for many detailed hints in the second edition.

devoted to systematisation, they contain besides a mass of special observations and of ideas going beyond these; and he influenced in a similar way a wide circle of pupils.[1] For example, at this very early period he already regarded the smallest living beings as the cause of infectious diseases, for which view he gives a number of pertinent reasons.[2] Particular mention should be made of the fact that Linnaeus, as a result of experience derived from his travels, and extensive breeding experiments on plants under the greatest possible change of conditions, soil, and climate, thoroughly satisfied himself concerning the astonishing invariability of existing species, but that he did not regard this invariability as unlimited. He observed the production of hybrid plants in the botanical gardens, and produced them experimentally; studied fertile and sterile hybrids, and saw in the production of such changed forms of plants a pointer towards a possible understanding of the production of new species in the course of long periods of time.[3] Furthermore, he regarded excessively favourable conditions of life as a special cause of the occurrence of new varieties. We see here how unjust it is that Linnaeus' great achievement of having demonstrated the very great and quite general stability of specific form, which is the foundation of all system in the plant and animal world, in later times gradually became almost a reproach to him, as though he had no desire to see any further; indeed, almost as if his personal characteristic had been the defence of complete invariability of species. Surely the fact alone

[1] Much of this is found in Linnaeus' *Philosophia Botanica* (Stockholm, 1751), and in the work published by Giesecke, *Prälectiones Caroli a Linné* (Hamburg, 1792).

[2] The many and varied achievements of Linnaeus are fully appreciated in detail in a work published by the Swedish Academy, *Carl von Linné's Bedeutung als Naturforscher und Arzt*, Jena, 1909. See particularly pp. 148 to 188 in Part IV.

[3] 'He dared to proclaim *plantas hybridas* and gave posterity a suggestion of *specierum causam*,' said Aszelius in a volume published by him in 1823, and containing notes and additions of manuscript jottings by Linnaeus concerning himself. German edition, 1826, p. 83.

JOHANN MENDEL

GUSTAV KIRCHOFF

Robert Bunsen

that he found good reasons for ranking *Homo sapiens* simply in the animal kingdom, was a particularly effective preparation for ideas which came later. Are we not to-day also still far behind Linnaeus, when we are unwilling to recognise the existence of human strains with their distinctive characteristics, and the necessity for selective breeding, although this is recognised in the case of all domestic animals?

A second great creator of order appeared a hundred years later, Charles Darwin namely; but he was also much more than a mere systematist. His great and original survey of all living things no longer starts from existing life as we know it at the present time, but rests upon the beginning of an understanding of the gradual development of life upon earth, a development upwards from the simplest to ever higher forms of living organisms. The view generally held and regarded as valid, which Darwin found when he began his work, in spite of Linnaeus' ideas having already reached further, regarded the different species of living creatures as having been set upon the earth from the beginning in their existing forms, and having then reproduced themselves from one generation to another in an identical form. However, petrified remains of animals had been found in excavations in the earth's crust, the forms of which were different from those of present-day animals, and it had to be concluded that these were of various ages, in accordance with the arrangement of the layers of rock one above the other. It thus appeared that in course of time certain forms died out, while new ones appeared, as though one had gradually taken the place of the other.

This gave a proof of the appearance of different species at successive times in the history of the earth, and hence almost of a gradual development of these species.[1] This fact

---

[1] Lamarck had already concluded from this fact, and for other reasons, in his *Philosophie zoologique* (Paris, 1809), that the assumption of the invariability of plant and animal species is untenable, and gave decided expression to this view.

suggested to Darwin the idea which showed the wonder of
the manifestations of life as connected with the still greater
marvel of the coming into existence of life on this planet.
His idea made it seem possible that these marvels could be
investigated by scientific methods, by methods therefore
which only strive after the determination of realities, and
which had already revealed so much that had appeared incomprehensible; though not, it is true, without having
shown behind every marvel thus explained a still greater
marvel, precisely in accordance with the position of man's
limited mind when confronted with the whole of nature.

The course of applying scientific research to these wonders
of nature was followed by Darwin with great persistence, by
collecting an unheard-of mass of relevant facts which, when
properly arranged, all pointed in one direction. They appeared to show that life on the earth had started from small
beginnings, and gradually developed in the course of the
long periods of time since the cooling of the earth's surface,
to ever higher forms and ever-increasing complexity; and
that this had occurred as the result of small variations appearing in the progeny of organisms, combined with the
effect of a continual selection of the fittest forms, namely
those forms best able to maintain themselves in the given
environment for further reproduction, and a continual destruction of unsuitable forms as a result of the perpetual
battle with the environment, which battle is actually a characteristic of all development of life. This very general idea is
developed by Darwin in every direction in his book *On the
Origin of Species*, which appeared in 1859. On account of
the indefatigable way in which he investigated everything
that told for and against the idea, and estimated it solely
from the point of view of truth, and because the idea thus
connected with a large range of facts is correspondingly fruitful, and has become of the highest value as regards further
progress in our knowledge of life, we have put forward

Darwin as an eminent investigator of life, in the same category as the great investigators whom we have considered from past times.

Charles Darwin, who came from a well-to-do landed family, was born at Shrewsbury in the West of England, as the fifth child of a physician.[1] From his earliest youth he was an enthusiastic observer of all kinds of living creatures, and showed also a lively interest for all sciences. When he was studying at the universities of Edinburgh and Cambridge, first medicine, then botany, and at his father's wish also theology, he made use of the opportunities of university life mainly in a free manner, and obtained by his own choice and responsibility a many-sided education. At the age of twenty-two, stimulated by the study of Humboldt's works, and with a desire to see the tropics, he made a voyage round the earth on a small vessel, which lasted for five years. From this he brought back so much experience, and such a mass of plant and animal collections, that he was occupied for several years in working over them. Thus began his very industrious, and from that point onwards, very retired life, first in a small house in London, and then permanently on a country estate, which he bought (having meantime married) and made into a fine country seat with garden, park, and hot-houses.

The idea soon came to him that it should be possible to find out something about the origin of species. After spending five years in collecting appropriate facts, he allowed himself to give them a thorough consideration, which led after two further years to the first definite conclusions, which he regarded as probable, and then subjected to continued further tests. Not until after a lapse of a further fifteen years was the work already named published. Many special papers and works by him which appeared before and after, containing a rich mass of new observations, also

[1] See *Life and Letters of Charles Darwin*, by his son Francis Darwin.

experiments in the breeding of plants and animals, and new ideas in these fields, thus standing in a nearer or more distant connection with the work we have named, can only be mentioned here without detail. Thus Darwin's life, in spite of poor health, was always filled with steady research, until he died at the age of seventy-three.

Like Newton, Watt, and Lord Kelvin, he was buried in Westminster Abbey in London. A particularly fine memorial to him is in the great hall of the Natural History Museum in London, where a mighty marble statue shows him seated amid the rich exhibits from the plant and animal world, the wonderful and astonishingly manifold forms of which he taught us to see in quite new connections, and with much profounder ideas, and which now appear to us as a large and closely connected family of blood relations, right back from the furthest ancestors, up to the human being who thus regards them. Darwin's work, after encountering at first very violent opposition, received the fullest recognition, which Darwin himself lived to experience. But there also followed great exaggeration, together with a new wave of materialism, inasmuch as it was made to appear, in a manner quite foreign to Darwin and his works, that nothing remained hidden any longer from our understanding.[1]

This may have been aided by the fact, that in Darwin's time the invariable validity of the energy principle, both for living and for lifeless matter, was practically assured, so

---

[1] Many may have been led astray respecting Darwin's nature by the fact that it was his lot at the close of his work on the origin of species, to find himself opposed by a generally held view of the complete invariability of species, and of their production by some such process of creation as is described in the Old Testament. Darwin was regarded by many as a destroyer of spiritual values, and his views were welcomed, and carried further, in this sense by some. Such complete misunderstanding of his work, as contrasted with the search after truth to which his whole life and all his writings bear witness, is again only a part of the thousand-year-old curse, which the Old Testament, regarded as a source of cultural fact and moral elevation, has brought over that part of humanity which seeks the light.

that those of a more superficial nature could now regard the whole world as a mechanism running according to known laws which only needed to be further investigated in a few details, and which now could be regarded as well-earned property open to the exploitation of enlightened mankind. But who, well acquainted with all known natural laws and yet not quite petrified in mind, can even see a plant in Spring, together with the coming and going of the flying insects which it attracts, and not recognise all this and everything that it suggests to our mind, as magical, and entirely unlike the dreary mechanism which such people imagine ? To gain further insight into life will certainly require great humility.

In actual fact it was again, as in the case of the great men of science of all kinds, a quiet devotion to nature, great patience, and a complete absence of all desire for fame, which again brought further progress after Darwin's time. The Augustinian monk Johann (called Gregor) Mendel united in himself all these qualities; he set to work at the point at which Darwin saw the greatest deficiency: 'the laws which govern heredity are completely unknown.'[1] Mendel opened the way to a knowledge of these laws by studies in heredity carried out on suitable living creatures, and himself attained important results, which could even be expressed in figures, and the prosecution of which still occupies a great many scientists.

We now know through Mendel, that the formation of hybrids, which takes place to such an extent, particularly in life when withdrawn from free nature, does not mean an inseparable mixture, or an obliteration of the qualities of the parents. Every living creature passes on unchanged to its descendants all the characteristics that it has itself inherited, though not all of them to every one of its descendants, since both parents are here concerned, although often in an invisible manner. New kinds of living creatures, also

[1] Darwin, *Origin of Species*, London, 1859, p. 13.

X3

new races, therefore do not arise by cross-breeding, the only result of which is to produce a mosaic-like combination of qualities, governed by certain laws, which combination may again fall apart. Suitably conducted selection by breeding can again remove from hybrids inherited qualities which are present, and can also again bring other qualities, when they have once existed, into the race again.

Mendel's results obtained by breeding experiments have been shown to be in most significant agreement with the results of microscopical investigation of the contents of the body cells and sex cells of living creatures. In this way, the study of the smallest living structure which can be seen under the microscope, the cell nucleus – and perhaps also that of the groups of atoms which compose it, and which cannot be seen separately – may be brought into connection with the study of the peculiarities of inheritance known to us in connection with the great cell-states animated by a single spirit, that is to say, the most highly developed living creatures, a study already forming the science of race. This would perhaps open up the possibility of approaching nearer to the great question of how we are to imagine, in a more satisfactory manner than has been possible since Darwin, and especially since Mendel, the coming into existence of inheritable characteristics of an entirely new description, not inherited from any ancestor, and yet transmissible, which must certainly have occurred in the course of the upward development of species upon earth. This question appears to be one which it will not be possible to answer upon the basis of the material world alone.

But these distant goals again will only be approached by natures of a kind as unprejudiced as Mendel. They will have to be men of science capable of assimilating facts of every kind which have become known, and then working them over like Darwin or Mendel, in seclusion, and adding to them, their attention far removed from those 'isms' in which small minds are accustomed to enclose the fragments of the achievements of great men, fragments which correspond to their own mental capacity. These investigators will have to go their own way in an entirely independent manner, led by their direct knowledge of facts.

Johann Mendel was born in a little village in what at that time was called Mähren, on the Silesian frontier, as the son of simple peasants. He was to take over the small piece of land owned by his father; but the familiarity he had gained with the life of plants awakened in the boy a desire for scientific study, which the parents only yielded to reluctantly, and were only able to fulfil by great self-sacrifice. The director of the Gymnasium, which Mendel was then allowed to attend, was an Augustinian priest; he no doubt led Mendel likewise to take the life of a monk, which allows of so much leisure for scientific activity, and mental retirement. At twenty-one years of age he was received into the Augustinian priory in Brünn, under the name of Gregor, after having finished his theological studies. He now had the possibility of studying botany, physics and chemistry at the university of Vienna, where Doppler was also one of his teachers. In the examination for teachers he failed, obviously on account of too great freedom of mind; but he was nevertheless able to act from the year 1854 onwards for fourteen years as an excellent teacher of natural history and physics at the school in Brünn. In these years also his most important experiments in breeding were performed, mostly on peas and beans, for which purpose the monastery garden was available. In 1868 he became abbot of his monastery; the duties thus undertaken, which he, like Copernicus in a similar case, fulfilled most faithfully under very difficult circumstances, made a premature end of his scientific investigation. He died in 1884, at the age of sixty-two.[1] The

[1] A detailed biography of Mendel has been written by Dr. Hugo Iltis (Berlin, 1924).

long postponed publication of his results only took place in great brevity and in a hardly accessible place;[1] they were therefore ignored for a long time. Only from the year 1900 onwards did the important and extraordinary significance of his discoveries become clear to everyone; and they now continue in their effect.[2]

ROBERT WILHELM BUNSEN (1811–1899)
GUSTAV KIRCHHOFF (1824–1887)

BUNSEN and Kirchhoff, from the days when spectrum analysis was a great and new conquest of the human mind, and old Heidelberg was the place sung by poets and singers, are a combination, which, in spite of all that has passed since their time, is still unforgotten, and will always remain so. Two rare men of science came together at the most fruitful time of their lives; they joined forces in the most fortunate manner for the purpose of investigating an entirely new road to the discovery of facts. This road led far beyond the known, and actually – such an event had hardly happened since Newton's time – beyond the earth and the solar system into cosmic space, to the discovery of the material composition (chemical analysis) of the most distant stars, concerning which, until then, only very slight changes of position measurable in the telescope were known, all questions of their

[1] Mendel's papers 'Versuche über Pflanzenhybriden' appeared in the *Verhandlungen des naturforschenden Vereins* in Brünn, 1866 and 1870; reprinted in Ostwald's *Klassiker*, 4th Ed., Leipzig, 1923.

[2] See W. Bateson's *Mendel's Principles of Heredity*, 4th Imp., Cambridge, 1930. Mendel's original papers are here translated in full. A well-known small book is Punnett's *Mendelism*, Cambridge, 1927.

Letters from Mendel (written between 1866 and 1873) which show with what difficulty his ideas were understood by specialists at the time, and how little, therefore, Darwin's warning quoted above was understood, even by those who supposed themselves to be adherents of Darwin's doctrines (Darwinism), have been published in the *Abhandlungen der Sächsischen Gesellschaft der Wissenschaften*, vol. 51, 1906, p. 187.

internal constitution being quite insoluble. Bunsen and Kirchhoff rendered this possible by taking up again the optical investigations of Newton and Fraunhofer, which in the previous forty-five years had hardly been advanced any further; they started from the question of the emission and absorption of light.

It was known that hot bodies, at a temperature below about 600° C, only emit the dark heat-rays, the infra-red discovered by Scheele, and that as the temperature rises, red light gradually appears (at a red heat), and finally the whole visible spectrum (white heat). Solid and liquid bodies exhibit in this respect, generally speaking, no great difference; any piece of glowing charcoal allows all these phenomena to be readily observed. Hot and glowing gases or vapours however, such as we find in flame, behave differently; at a sufficiently high temperature they are able to glow with coloured instead of with white light, the colours depending not so much upon the temperature, but rather upon the material nature of the gas or vapour. This was long known in pyrotechnics, and used for the production in fireworks of flames of almost any desired colour, for which purpose a large number of recipes always existed; it was further also known that the flame of alcohol, which in itself emits little light, becomes capable of emitting coloured light when certain substances are brought into it. A yellow colour in particular always readily appeared in the alcohol flame, but often without any recognisable cause, so that there was a certain doubt as to what the colour was due to. What was wanted were clear experiments, which would eliminate, with certainty, all matters of chance in order to allow the true causes of the effects observed to be recognised. When this had been attained in the matter of these flame colouration, they might, for example, also be available for indicating the presence of certain substances.

Bunsen was the true investigator for such problems; he

was never satisfied with what was unclear or half done, and at the same time he had a limitless inventive power for creating new means of satisfying his need for thorough investigation, and was quite tireless in following ways, no matter how difficult, which led to a clearly seen goal. His investigations of coloured flames therefore led to a great deal more than merely to chemical analysis by means of flame colourations, and not least because they were happily combined with Kirchhoff's insight into everything that had been ascertained from the time of Newton and Fraunhofer to Clausius.

For clean experiments with flame, the first necessity was a cleanly burning flame; the alcohol flame with its wick, which caused all kinds of disturbance, was not a flame of this kind. Also for many other laboratory purposes, a flame was already much needed, the properties of which could be regulated by hand in a trustworthy manner. Illuminating gas, which had already been used for some fifty years in England, appeared much more suitable from the start than alcohol or other liquid fuel; but Bunsen nevertheless was not in the least satisfied with the English gas burners made for laboratory purposes. When, in 1855, gas lighting was introduced in Heidelberg, Bunsen immediately set to work to invent a suitable burner; the Bunsen burner at once became quite indispensable, and has remained so till to-day, when it is known to everyone.

But the result was not only the sole source of heat suitable for the laboratory, and a fundamental constituent of all technical heating arrangements, but above all a flame of hitherto unheard of steadiness, and at the same time versatile in its properties. But none but Bunsen himself understood how to make full use of it for his remarkable flame reactions – a new chemical test for all kinds of substances, which could otherwise only be investigated by the use of wet reagents, or at least by the blowpipe – and by making use of the cleanliness of the experimental conditions, which this flame allowed of, he was able to lay the foundation of spectrum analysis.

It was now possible to evaporate on a pure platinum wire any substances whatever in the non-luminous gas flame, and so cause them to emit light, and it was now a promising step to continue to purify the substances used by every means known to chemical science, until each of them developed its peculiar flame colour fully and unmixed, a task which Bunsen undertook with admirable persistence. In the course of this work, however, the flame colouration was not observed only with the naked eye, but also, following the methods of Newton and Fraunhofer, in a state of spectral decomposition by means of a prism. It then appeared, that the yellow colouration of the flame, so often observed, and having a yellow line already well-known to Fraunhofer, belongs exclusively to sodium, and that the uncertainty hitherto felt with regard to the origin of this yellow light was on the one hand to be referred to the incredible minuteness of the quantity necessary to produce this colouration quite visibly, or to exhibit the spectral line, and on the other hand, to the universal distribution of sodium in the form at least of traces as an impurity in practically all substances found on the surface of the earth. But we now had a means of recognition of the constituents of all substances which could be obtained in flames, or in electric sparks, or arcs, or otherwise in the state of glowing gases. This means allowed the certain recognition of the smallest possible quantities, while being easily and simply applicable; spectrum analysis by emission was thus founded.

This led Bunsen himself very quickly – five and six years after the introduction of his burner – to the discovery of two new elements, rubidium and caesium, which had hitherto been unrecognised on account of the fact that they only occur in traces, as a rule, and are very similar to the well-known alkali metals. This spectrum analysis was just as well suited to the discovery of new elementary substances, as to proving the presence of already known elements, since almost all substances are decomposed, when in the state of glowing gas, into their elementary components, so that the single atoms occur in a free condition, and are able to exhibit the emission of light peculiar to them alone. So in the following years, other elements such as thallium, indium, gallium, scandium, germanium, were discovered, and the separation and certain recognition of others, such as the noble gases, were rendered possible. The result was a very great enlargement of our knowledge of the elementary constituents of matter. Thenceforth one quite indubitable proof of the existence of a new element was required: it had to show new spectral lines.

Kirchhoff's own particular addition to spectrum analysis, and one of equal magnitude, related to the absorption of light. He not only recognised that this stands in the closest relation to emission, but he also gave a proof of this fact, placing the connection among the best founded parts of scientific knowledge, and giving it the clear statement which is essential if such knowledge is to have its full value (1860); this is known to-day as Kirchhoff's law. Kirchhoff's proof of the law consists firstly of a series of imaginary experiments of an allowable description, that is employing only means and processes capable of being realised with sufficient approximation, and secondly of the corresponding calculations, which allow the results of these experiments to be connected in a completely satisfactory manner with Clausius' second law of thermo-dynamics, in particular with the general fact, that heat never passes of itself from a colder to a hotter body.[1]

[1] The corresponding calculations occupy a very large space in Kirchhoff's statement. By suitable change in the imaginary experiments, they could be very much reduced in extent without injury to the convincing value of the proof. As regards our estimate of Kirchhoff's achievement (which he himself, in face of objection made to it, defended satisfactorily; see

Kirchhoff's law thus stands together with these facts of experience, and with the mass of other experimental facts collected together in the theory of heat, and is thus itself as well secured as all these experimental facts themselves. The law states that the emission and absorption of light of definite wave-length are proportional to one another in the case of all bodies at the same temperature. A body – such as sodium vapour, for example – that emits the yellow light corresponding to a definite position in the spectrum, and so of definite wave-length, will also absorb the same light to a corresponding degree, and if it does not emit other kinds of light, it will also not absorb them. This means that a flame fed with sodium and placed between a very bright white source of light and the prism apparatus, must likewise cause a dark line to appear in the same position as the bright line given by the flame. Kirchhoff himself observed that this is actually the case. This provided a special experimental foundation to the law which in Kirchhoff's case preceded its proof.

But at the same time an artificial Fraunhofer line had been produced in the spectrum, at exactly the same position as the line named the D line by Fraunhofer in the spectrum of

Gesammelte Abhandlungen, p. 573) this remark is of no importance, but the same is not true regarding the fact that all such proofs of natural laws have no more and no less value than that of demonstrating convincingly a connection between the new laws to be proved and one or more other laws already regarded as certain, this connection being of such a kind that the new and old laws now stand or fall together, if nature is to remain true to itself (which assumption has always been found to be justified in all non-living matter). In so far as the old laws made use of have been ascertained by experience to hold, that is to say to agree with reality, the same must also be true, in view of the proof, as regards the new law. It is a matter of complete indifference how much or how little calculation is made use of in constructing the connection in question. If there is much calculation, and if at the same time the laws of experience made use of are not very clearly set forth, we get the deceptive appearance of a 'mathematical proof,' which appearance obscures and hides the high character-forming value of scientific investigation, by leaning to the opinion that its true content is only accessible to mathematicians, or that mathematics of itself leads to knowledge of nature.

the sun. According to the law, which is assured by so many other general experimental results linked to it, this cannot be a matter of chance; it is necessary to assume that somewhere in the path of the ray of light from its origin in the sun to our spectroscopes on the earth, sodium vapour exists, consisting of free sodium atoms as in the case of the flame. But this can only be in a gaseous atmosphere at a high temperature on the sun itself, an atmosphere which must be traversed by the white light radiated from an incandescent liquid or solid core of the sun. In this way a well-founded explanation of the hitherto unexplained Fraunhofer lines in the sun's spectrum was given, and furthermore, by the aid of Bunsen's results, the possibility of a chemical investigation of the sun's atmosphere was opened up.

Few investigators have ever been able to rejoice at one time in so many important results as Bunsen and Kirchhoff in this case.[1]

Kirchhoff now measured the dark lines of the sun's spectrum still more elaborately than Fraunhofer had already done, and he likewise measured the emission lines of elements, particularly iron, in order to be able to determine coincidences in full detail. In this way, besides sodium and iron, hydrogen, magnesium, calcium, and several other elements known on the earth were proved to exist in the sun's atmosphere. In the case of one of the dark lines in the sun's spectrum, the element to which it belonged was only subsequently found upon the earth, as giving a bright line in the same position; this element was therefore called helium (one of the noble gases). At total eclipses of the sun, when the centre of the sun is obscured by the moon, the bright spectrum lines are also seen – for the first time in the year

[1] The place in which this discovery was made is indicated on the house called 'Im Riesen' in the Hauptstrasse in Heidelberg by means of a tablet. On the opposite side of the road is the Friedrichsbau, built three years later, into which Kirchhoff then moved; the chemical laboratory on the Wredeplatz had already (1855) been previously rebuilt for Bunsen.

1868 – as coming from the sun's atmosphere, which then alone is giving light.

In this way the 'protuberances' of the sun were recognised as stupendous outbreaks of glowing hydrogen gas, and it was then found possible to examine them thoroughly by means of the spectroscope even in the absence of an eclipse. Attention was then paid to the spectroscopic decomposition of the light from the fixed stars and nebulae, in which matter Fraunhofer had already made a beginning; it was now possible to perform the same chemical analysis, which had already succeeded in the case of the sun, upon these bodies, which for the most part are many thousand and more light-years distant from us.

It was shown that the whole visible universe, even to the greatest observable distance, consists of no other materials than those found upon the earth. Matter is everywhere alike in character; the universe is a whole; the enormous intervening spaces between one solar system and another do not involve complete separation; all that is familiar to us on earth concerning matter holds good also to the limits of the universe.

To these achievements of spectrum analysis further results were soon added, which also concerned the processes taking place in the heavens. Christian Doppler, born in 1803, son of a master-mason of Salzburg, and professor of mathematics and physics at various institutions in Austria, finally at the university of Vienna, did not live to learn the results of Bunsen's and Kirchhoff's work; he died early, at the age of fifty. But he had discovered a fact which could now be made good use of: 'Doppler's principle.' This relates to all propagation of waves, including light, and states what happens when the source of the wave, or the observer, move in the direction of propagation. If the motion is not rapid as compared with the rate of propagation of the waves, all that matters is the change in the distance between the source and the observer; and the principle states that when the two approach one another, a shortening of the wave-length seems to the observer to occur, while a lengthening takes place when they recede from one another; the two effects being in a definite dependence upon the rate of change of the distance apart.

The principle was first experimentally proved in connection with sound waves, the source of sound being situated upon a locomotive in rapid motion; the pitch of the note appears to the observer raised when the source of sound is approaching him, and lowered when it is receding from him, as the principle requires. In the case of light we have to expect, in place of a change of pitch, a displacement in the position on the spectrum, when the source of light and the observer change their distance from one another. The rate of change must, on account of the great velocity of light, be considerable, if the displacement in the spectrum is to be capable of observation. For this reason it was for the moment only to be expected in the case of the heavenly bodies, in which great velocities frequently occur. Observations capable of clear interpretation became possible as soon as the lines in the spectrum of the stars had been assigned by Kirchhoff and Bunsen with certainty to definite elements.

The position of every definite line in the spectrum of a certain star could then be compared with the position of the same line of the corresponding element in an earthly source of light; from the distance apart of the two lines the rate of change of the distance between star and earth could then be calculated by Doppler's principle. We already know these 'radial velocities' in the case of thousands of stars, which means for us an entirely new possibility of discovering the actual motions taking place in cosmic space. Also in the case of the internal motion of certain heavenly bodies such as the nebulae, spectroscopic observation of the radial velocity has already led to important conclusions.

Special mention should be made of the discovery of many

'spectroscopic double stars,' which even through the best telescopes can only be seen as single stars, since they are much too far away. But the regularly recurring doubling of the spectral lines of these stars tells us that, and how quickly, one of the two suns revolving about their common centre of gravity approaches us, while the other recedes; and from this we are able to calculate the orbit and in many cases even the masses of these double suns in a logically consistent manner. An important direct confirmation of the validity of Doppler's principle for light has finally resulted from the observation of earthly sources of light (glowing atoms in the canal rays) moving at extremely high speeds.

Spectrum analysis with all that has been built upon it, by no means forms the only remarkable achievement of Bunsen and Kirchhoff.

Regarding Kirchhoff, we may mention his laws concerning the flow of electric current in networks, and in conductors of two-dimensional or solid form; these being enlargements of Ohm's law. Bunsen further advanced science by a large number of other researches which can only be mentioned here in part, but concerned almost all directions of investigation, and everywhere pointed out new roads and provided new methods of attack. Everything that he undertook was given new shape by him, and at the same time generally became of practical importance. Thus he discovered, continuing Davy's electrolytic researches, the 'Bunsen element'[1] which then remained for a quarter of a century the only existing useful source of strong electric current.

By its use he then found the best method for the electrolytic preparation of pure metals such as calcium, aluminium, magnesium; all of these became of technical importance later. Burning magnesium, and also Davy's arc light fed

---

[1] This contains zinc in dilute sulphuric acid, and gas retort carbon in concentrated nitric acid, and dates from the year 1841; later (1875) Bunsen also suggested the chromic acid cell, which is likewise much used.

with the strong currents from his new elements, provided him with sources of light of hitherto unattained intensity, and for the purpose of measuring them he invented the grease-spot photometer, which for fifty years was the only photometer used. A later refinement was to replace the grease-spot by a special cube of glass.

His preparation of new elements in a pure state led him to invent the ice calorimeter, in order to be able to measure, for the purpose of atomic weight determination, the specific heats of the elements with great accuracy, in spite of the small quantities available. This calorimeter again provided the means for further investigations, and remained for a very long time unequalled by any other method.

Bunsen had also not a little to do with investigations concerning the interior of the earth, volcanic phenomena, and the formation of rocks. On a scientific journey to Iceland (1846) he discovered the nature of geysers, and later investigated the dependence of the melting-point of materials upon pressure, most substances being specifically heavier in the solid than in the liquid state. He found that the melting-point rose rapidly as the pressure increased, and drew attention to the high pressures which undoubtedly exist in the interior of the earth, where we must therefore assume a considerably increased melting-point for rocks and substances composing them. This agrees with the fact that the interior of the earth must be for the most part solid, as already follows from Newton's investigation of the phenomena of tides, inasmuch as the body of the earth only takes part in tidal motion to a very small extent, a fact which cannot be ascribed solely to the solid state of the surface. Furthermore, Bunsen's results concerning melting-points formed a valuable confirmation of a calculation which at that time had just been based upon the thermo-dynamics of Clausius and W. Thomson. This same calculation led the opposite kind of behaviour to be expected in the case of ice, which is specifically lighter than water: the melting-point should fall as the pressure rises.   This was then tested and found to be actually the case.

Bunsen's work on cacodyl and its compounds was also important.   He recognised in this substance, consisting of carbon, hydrogen, and arsenic in fixed proportions, and first prepared and examined by him, a 'radicle,' that is to say a group of atoms always remaining together, which behaves like an element, and hence forms combinations of various kinds.   More than twenty years earlier, Gay-Lussac had already discovered another radicle, cyanogen, consisting of one atom of carbon and one of nitrogen, and forming a first and particularly simple example of its class; cacodyl was an example of a very complicated radicle, since it contained altogether nine atoms, and led us again a step further in our knowledge of the polyatomic substances of organic chemistry.   The investigation of the cacodyl series may have particularly appealed to Bunsen's temperament; it was a challenge to his skill to master substances which were explosive, self-igniting, dangerous to life as poisons, and of an unendurable odour even in traces.   He had already previously found for the poison arsenic an absolutely certain antidote – at that time a very important matter, since arsenic was one of the commonest poisons.

It is noteworthy that Bunsen, in his publications concerning the cacodyl series, which belong to his youth (1837–1841), makes use of the same atomic weights for representing the composition of the compounds discovered by him, as are known to be correct to-day (according to which, for example, water is written H_2O); but he later abandoned this and took for example the atomic weight – or as he preferred to say 'combining weight' – of oxygen as only eight times as great as hydrogen (instead of sixteen times), as was the assumption at a time when Avogadro's law was still unknown or only an hypothesis; this leads to water being written HO, as familiar to us in Faraday's work. Bunsen thus withdrew at a period of uncertainty from the quarrel about 'hypotheses which are changeable, and are often changed.'[1]

By means of the combining weights, which were free from hypothesis, he was just as well able to express in formulae the actually observed composition by weight of all compounds, as by using the atomic weights, the true values of which could only be discovered beyond doubt later, after Avogadro's law had been rendered certain by Clausius' proof of the kinetic theory of gases. When this matter had been cleared up, Bunsen refused to make a further change, and adhered to combining weights. This, however, prevented him from penetrating into the manner in which the molecules of compounds are built up from their atoms, since this requires the use of formulae founded upon true atomic and molecular weights. This advance, which represented an important step forward, was left to his younger contemporaries and successors to accomplish. In effecting it they were able to rely upon Avogadro's law, which was already proved to be true, and gives the correct molecular weight. It then appeared that we must ascribe to the various atoms various values expressing their power of combination with other atoms, according as, for example, they are able to combine with one, two, three or more hydrogen atoms, to form a molecule; this value for the combining power also holds good in general as regards the combination of the various atoms among themselves. It is thus possible to ascribe to each atom, in proportion to its combining power – its 'valency' – a corresponding number of valency positions, and to assume that when a molecule is formed, atoms are linked together because their valencies 'saturate' one another in pairs.

These hypotheses agreed with Davy's and Berzelius' ideas concerning the electrical nature of the forces holding

---

[1] Bunsen thus expressed in his lectures his decided distaste for hypotheses which are often put forward as finished theories.

atoms together in combination, and with Faraday's second electrolytic law, to a certain extent, but not in every respect. The fact that they could not satisfy Bunsen, particularly when they received an all too definite expression in 'structural formulae,' is still more self-evident than in his time; for in spite of the fact that essentially new data have since come to light, nothing final is yet known concerning the structure of molecules. Nevertheless, the idea of saturation, which was expressed in the formulae in question, has become extremely valuable in correspondence with its content of actual fact; particularly as regards the numerous organic compounds, it has become, since the introduction of Kekulé's benzine formula (1865), the best means of orientation and guide to work as regards the preparation of large numbers of these compounds.[1]

The life stories of both Bunsen and Kirchhoff were of a very simple character.

Robert Wilhelm Bunsen was born in Göttingen as the son of a professor of philology; he entered the university of his birthplace at the age of seventeen, and later became a member of its teaching staff after visits to Berlin, Paris, and Vienna. In the year 1836 he was called to the technical school at Cassel, two years later to the University of Marburg, then to Breslau, and soon afterwards, in 1852, to Heidelberg. Here he remained for thirty-seven years, and became one of the most eminent figures of university life.

Famed both as a discoverer and teacher, he always drew a large number of pupils to his laboratory; scarcely any scientist of the next generation and none of the founders of the German chemical industry which soon developed, but worked at least a few terms with him. His experimental lectures were unique in their elegance, in which Bunsen's

---

[1] August Kekulé lived between 1829 and 1896; he was born in Darmstadt, became assistant under Bunsen, and finally professor of Chemistry in Bonn.

Ys

whole personality was reflected, and at the same time in their natural simplicity, combined with perfection of form; the joy of the discoverer was communicated to the hearer in all his demonstrations and explanations.[1] He was a shining example of the cultivation of pure science; the idea of gaining personal advantage from his inventions, which would have been easy in the case of the burner, the photometer, and many others, was entirely strange to him. His life lay almost entirely in the laboratory among his pupils, who were also able to be witnesses of his own researches, as they were being carried on. His remarks were often full of humour under the mask of the greatest seriousness; his mode of thought produced its impression without making any direct claim to superiority. He did not found a family; in the university vacations he loved to make long journeys, generally to the south, often accompanied by friends among his colleagues.

At the age of seventy-eight Bunsen retired from teaching; for a further ten years he enjoyed walks, and later drives, in the woods around his beloved town, from which no offers, however attractive, had ever been able to separate him.

Gustav Kirchhoff was born at Königsberg in Prussia as the third son of a legal dignitary, and also studied there. He entered the University of Berlin in the year 1848, and then became an assistant professor in Breslau. There he first met Bunsen, who in 1854 obtained an invitation to Heidelberg for him. In the year 1875 Kirchhoff left this scene of his most successful activity, where he had also married a second time after the early death of his wife, and went to Berlin, to work alongside Helmholtz. An injury to his foot caused by a fall on a staircase, had been cured, but increasing infirmity appeared, and he died twelve years later at the age of sixty-three.

---

[1] Wider circles also knew of them; princes and men eminent in intellectual life who travelled through Heidelberg were not seldom seen in his lecture theatre – which lay at the corner of Academiestrasse and Plock – among his own students.

JAMES CLERK MAXWELL

1831–1879

CLERK MAXWELL summed up our whole knowledge of the ether as far as it went in his time, and further made so fortunate a use of knowledge then existing only in the form of general indications, that a well finished structure resulted, which – as experience later showed – really corresponds in a very wide degree to the facts. This structure has a mathematical form; it consists of certain equations, which state quantitative connections between the states of the ether in space, measurable in the form of electrical and magnetic forces, and the constants representing the properties of the matter likewise present in space.[1]

Along with these appears, as a constant of the ether, the velocity of light. The equations are differential equations; that is to say, they relate only to a volume element of space and to its immediate neighbourhood, and also only to elements of time. This entirely corresponds to Faraday's idea, derived from all his experience, that it is not forces at a distance – which leap over all volume elements – that are active in the ether, but that everything occurring in it acts only from point to point.[2] Altogether, Maxwell founded his equations entirely upon Faraday's conceptions. He summarised in them everything fundamentally known concerning light, electricity and magnetism, that is to say, all that taken together, constituted the knowledge of ether at that time. The fact that he was able to include the phenomena of light, which had been since Huygens' time the starting point of the physics of the ether, although only

---

[1] These material constants are: (1) the dielectric constant or specific inductive capacity introduced by Faraday; (2) a constant analogous to the latter, but referring to the magnetic force – the magnetic permeability; (3) the electric conductivity.

[2] Wilhelm Weber likewise carried out a magnificent summary, but one founded on the assumption of action at a distance; facts later showed that it did not correspond with reality.

electric and magnetic forces appear in the equations, was also derived from a thought of Faraday's, according to which light and electro-magnetic forces might 'by no means impossibly' be phenomena of one and the same ether.[1] This idea of the unity of the ether became the great fundamental idea in Maxwell's theory as expressed by his equations. The fact that it is a theory (summary of knowledge), and not simply an hypothesis (supposition), was only somewhat dimly indicated in Maxwell's time by Faraday's discovery of the electro-magnetic rotation of the plane of polarisation of light and – more obviously – by the appearance, as the result of Wilhelm Weber's researches, of the velocity of light in electrical phenomena and measurement.

Maxwell's equations summarise, in a manner astonishing even to the mathematician, the fundamental results of whole series of researches. They contain Coulomb's two laws with the essentials of the potential theory, Oersted's discovery and Ampère's researches, Ohm's law, Faraday's law of induction; all this being supported on Faraday's conception of lines of force, which is thus for the first time shown to be in full harmony with the older potential theory of Laplace. They further contain – as a hypothetical addition made by Maxwell – the existence of electric waves, which, starting from the electric oscillations already treated of by W. Thomson (Lord Kelvin), are propagated in free space with the velocity of light, are transverse like light waves, and should be refrangible like the latter in material media. So magnificent a summary in a few equations had hitherto never been known.

As regards their power of application, the sense of the equations, in which besides the Cartesian co-ordinates of space, time appears as an independent variable, is as follows: let the electric and magnetic forces at an initial moment

[1] Tyndall collected several remarks of Faraday's in this connection, in *Faraday as a Discoverer*, London, 1870, p. 154 ff.

everywhere in space be given; the equations then allow these forces to be calculated, by means of the rules of mathematics, for all future time; and therefore likewise every consequence of the given initial condition according to the laws summarised. We notice the similarity in structure and applicability between Maxwell's equations, and, for example, the fundamental equations of hydrodynamics. The latter also are differential equations, that is they relate also only to volume elements of space and their immediate neighbourhood, and they allow – as far as the powers of mathematics go – the prediction of the future state of the liquid from its initial state. What, therefore, hydrodynamic equations do for a liquid, Maxwell's equations do for the ether. Both sets of equations only depend upon experience; but while the hydrodynamic equations contain in the main nothing but Galileo's and Newton's laws of motion, the content of Maxwell's equations is, as we have explained, very complex; curiously enough, without their being unduly complicated. This fact, as well as the simplicity of the fundamental concept of the unity of the ether, already tells in favour of their agreement with reality, that is to say, for the correctness of the added hypothetical matter. It was only the inaccessibility of the ether which, in this case, made it necessary for many separate and partial discoveries to be made, which could then be simply combined; while in the case of liquids, which are more tangible, it was possible immediately to arrive at the final goal. The agreement with reality was proved fifteen years later by Hertz's discovery and exact investigation of the hypothetical electric waves.

It is noteworthy that Maxwell did not arrive at his equations directly by mathematical summarisation, but – starting from Faraday's experimental investigations – by a thorough consideration of possible mechanical ethers, which might give the observed effects which were to be summarised. He later

abandoned these mechanisms almost entirely, and only retained the equations, whereby he reverted completely to the
mode of thought of Faraday, who regarded his lines of force
simply as pictures of unknown states of a medium filling
space.   Hence Maxwell himself says that he has put Faraday's ideas into a mathematical form.   In the two volumes
of his book *On Electricity and Magnetism* which appeared in
1873, he by no means put the special results of his own differential equations in the forefront, but developed the then
existing body of knowledge of electricity and magnetism in
such a way that the statement of it is connected as far as
possible with Faraday's conceptions, and hence with his own
equations; which, however, only appear towards the end of
the work and are almost hidden from sight.   The special
features of Maxwell's work could only take full effect after
their experimental confirmation by Hertz.[1]

Nevertheless, even before this, Maxwell's work resulted in
Faraday's idea of lines of force being everywhere accepted,
and hence also by technical workers.   This was all the more
important, since Faraday's induction had found increasing
application in the production of strong electric currents,
particularly for lighting purposes; for the 'dynamo-electric'
principle of Siemens had shown us how to replace the
inconstant steel magnets of 'magneto-electric' machines by
electro-magnets, which were increased in strength by the
induced current itself (1860).   What then was still wanting,
as compared with the present-day perfection of dynamos and
electric motors, was the means of finding the best shape for
the iron and copper circuits of these machines; and the way
to this, simple but long overlooked, was finally indicated by
Faraday's line of thought.   The unnecessarily long limbs of
the electro-magnets disappeared, and the modern compact

---

[1] Hertz was also the first who was able to express clearly these special
features, together with the equations, free from all superfluous matter,
after the natural phenomena themselves had shown the way.

form, most favourable to the development of lines of force, brought with it the high efficiency customary to-day in the transformation of mechanical into electrical energy, and vice versa. The foundations of electrical engineering were laid.

Maxwell in his short life had many other achievements to his credit; of these we can only mention here his contributions to the kinetic theory of gases, which were directed towards deeper penetration into the subject, and an extension of Clausius' work.

James Clerk Maxwell, born in Edinburgh, came from a very old Scotch family. He was educated at the family seat, entered at the age of thirteen the university of his birthplace, and went three years later to Cambridge. After he had distinguished himself by a series of geometrical, mathematical, and optical publications, and also some concerning lines of force, he was made professor of physics in Aberdeen in 1856. Two years later he married. From 1860 to 1865 he was Professor of Physics at King's College in London: he there met Faraday. Later he retired to his estate, where he lived, apart from a journey to Italy, and short visits to London, entirely for his scientific work, which then resulted in his great work on electricity and magnetism. In the year 1871 he again decided to accept a professorship; this was in Cambridge, where, as a novelty, a special physical laboratory was built, which he opened. But only eight years later, in his forty-eighth year, a sudden illness put an end to his activities. The important confirmation of his mathematical and theoretical structure by the discoveries of Hertz was made after his death.

WILHELM HITTORF (*1824–1914*)
WILLIAM CROOKES (*1832–1919*)

Just as Galvani, eighty years previously, had been able to draw forth a new phenomenon from the inexhaustible riches of nature, a phenomenon which, not again forgotten, led to the development of quite new and undreamed of fields of knowledge, so did Hittorf and Crookes also follow up the investigation of processes which, up till their time, being outside the frame of science as ordinarily taught, had been simply regarded as curiosities, which could not otherwise be investigated. The phenomena were those connected with the discharge of electricity through rarefied gases. The processes had been known since Guericke invented the air pump and the electrical machine; Faraday, who later produced in the induction apparatus a still more suitable source of high tension electricity, had also observed these extremely curious phenomena, with their beauties of form and colour, which fill a space in which the pressure of the air is sufficiently reduced, when an electric discharge passes through it. It was also readily possible to follow the gradual transition from the simple short sparks exhibited at atmospheric pressure – which Leibniz was the first to see – to the extensive and complex phenomena which occur as the air pressure is gradually reduced. However, the maze of phenomena which occurred gave no indications of a way which might lead to an understanding of them.

Hittorf and Crookes were the first to find this way, by selecting from the mass of interlinked but incomprehensible processes a single one, and making it the subject of a special investigation, the example chosen being one distinguished by a certain simplicity. This was a process connected with the negative pole of the discharge – the cathode, as we say here as well as in the case of electrolytes – which had the effect of a radiation proceeding in straight lines from the surface of this cathode: the 'cathode rays' discovered by Hittorf. Hittorf was ten years ahead of Crookes in this investigation. But Crookes was able to follow the way thus indicated as the first, and for the time, only understanding mind after Hittorf,[1] and he was able to obtain better and newer experimental apparatus, in particular air pumps and other technical assistance. He showed his power of understanding by making very great further progress in the direction of solving the peculiar character of the phenomena of radiation, so that from this point enough was known to lead investigators of nature almost irresistibly in a certain direction, namely to the pure observation of these rays. Out of this hardly less actually developed – although a further fifteen years were required for the purpose – than Crookes had promised when he said: 'Here we shall find ultimate truth.'

Crookes and Hittorf were in every respect excellent complements of one another in this fundamental research. While Hittorf was very careful not to go beyond facts for which he could vouch with certainty, Crookes opened up vistas with the joy of the discoverer, and communicated this joy to others. Although many of his views were erroneous, his faith in the richness of nature was justified, and, holding it so strongly, he was able to induce successors gifted with equal confidence and ability to take the correct direction with complete certainty, when they set out on their discoveries. Hittorf's publications were dry, and almost repulsive; their great value was buried in the depths. Crookes announced with enthusiasm what he observed; but he had in England the best opportunity for doing so.[2]

[1] There had been many observers of discharge phenomena before and after Hittorf; it was a field of work which easily led to descriptions without end of what had been seen. But all these descriptions only led, in the best case, to real new knowledge, when that lacking in them all, namely the attempt to reach understanding by means of clearly devised experiments, had been supplied from another side.

[2] Hittorf published his results 'On the conduction of electricity by gases' in Poggendorf's *Annalen*, Vol. 136, 1869. Crookes gave in 1879

Wilhelm Hittorf was born in Bonn, where his father was a merchant. He studied in his native town with a short interval in Berlin, and gained his doctorate at the age of twenty-two. A year later he became a member of the staff of the university of his native town, presenting as his thesis an experimental research to which he had been led by his own observations on electrolysis. At the same time he also received a call to the Academy (later university) at Münster in Westphalia, as a teacher of physics and chemistry. It was very fortunate that he thus obtained an independent sphere of action at so early an age. However, he then remained for the whole of his active life in the same position, although his external circumstances improved; he never received a chair at any of the greater German universities. His fate was much the same as that of Ohm.[1]

Hittorf however was even more reserved by nature than Ohm; but since the many confidential petitions to ministries and princes made by the latter had no noticeable effect in giving him a more advantageous opportunity for work, Hittorf's reserve cannot have made much difference. At the age of sixty-five Hittorf retired from his position of teacher in Münster. He lived in his own house and garden, his younger and also unmarried sister keeping house for him. He died at the age of ninety.

William Crookes was born in London and also lived there for the greater part of his life. He appears to have educated himself mainly by the study of literature, and by his own experimenting. At the early age of twenty-three he obtained

in the Royal Institution a lecture entitled 'Radiant Matter, or the Fourth State of Aggregation, and this became still better known – also in its German translation – than his publication in the *Proceedings of the Royal Society*.

[1] Both of them received distinctions and honours in their old age; that however is of no importance, for what matters is to be recognised, and to have opportunity to work when still young enough. The fact that no specialist of influence in Germany took any notice of Hittorf became particularly clear when Crookes's later achievements were the subject of general admiration: Hittorf still remained un-named without protest.

a post as a chemist, which allowed him to found a family. His important work was done in his own laboratory, which was much less cramped than Hittorf's university laboratory. Crookes reached the high age of eighty-seven.

From the work of Hittorf and Crookes, large sections of the physics of the ether have subsequently been developed. It was only necessary to carry out exact experiments with Hittorf's glow discharge, and Crookes' 'radiant matter' or cathode rays, in such a way that they could be observed under more various conditions than had hitherto been the case in the discharge tube. This happened fifteen years after Crookes' publication (1894). From that point all the further progress followed quickly, which can only here be indicated in outline: the discovery of high frequency rays (X-rays), of radio-activity and radium, as well as of the other radio-active elements of high atomic weight and their radiations; the assured explanation of the nature of cathode rays, and then also of the rays of the radio-active elements, and of allied radiation such as the canal rays; the conduction of electricity in gases and finally, in particular, the phenomena of discharge at low pressures – which had been the starting point of the whole matter – became fully understood. The cathode rays themselves turned out to be free negative elementary electric charges separated from atoms – they were called 'electrons' – just as Wilhelm Weber had imagined them, without however, having been able to say anything concerning their actual existence, and still less concerning the possibility of completely separating them from matter.

Also, the rays of Hittorf and Crookes brought quite fresh information concerning the atoms of matter themselves. This agreed in essentials with that already furnished by Clausius from the kinetic theory of gases, but went beyond it; the interior of the atoms now became accessible to investigation, in spite of their smallness, which removes them from direct access by our senses. All atoms proved to be built up of positive and negative elementary charges of electricity in a manner which is still the subject of further investigation. It is easily comprehensible that so great and rapid an enlargement of our knowledge (it took place mainly in the years between 1894 and 1905) brought with it many new possibilities – which also included technical, and particularly at first, medical applications – but that it has also produced some confusion, all of which is of great influence at the present time.

Pre-eminent among the recent results of the further investigation of atoms is the fact that they can be split up; this can be effected by the action of very rapidly moving helium atoms, such as are furnished by the X-rays. This renders possible the transformation of heavy atoms into lighter ones by human agency, a transformation which occurs of itself in the case of radio-active elements. This, however, by no means exhausts the work and scientific legacy of Hittorf and Crookes; we must still add the following.

Hittorf was the first to continue Faraday's work on electrolysis to an important extent. He made a careful quantitative investigation of the motion of the ions in electrolytes during passage of the current, which opened up the road to a knowledge of the internal constitution of electrolytes, such for example as dilute salt solutions. Friedrich Kohlrausch began his investigations at this point and obtained further success, upon which depends our present-day exact knowledge of electrolytes, and the processes taking place in them.[1]

[1] Friedrich Kohlrausch, 1840–1910, was born at Rinteln on the Weser. His father, who was a teacher at the High School, had carried out, with W. Weber, the important measurements which led to the velocity of light being found as the proportionality factor for purely electrical quantities; it was also the father who induced the son to study physics with Weber in Göttingen. Friedrich Kohlrausch also later became a member of the staff of this university, and then assistant professor. In the year 1871 he was called to the Technical High School in Darmstadt, in 1875

Hittorf also began the thorough investigation of the conduction of electricity by flame, for which purpose he was able to make use of the Bunsen burner for exact experimenting.

Crookes was the first who attempted to follow up experimentally the pressure of light given by Maxwell's equation. This pressure, exerted by light or ether wave radiation on any surface which it meets, had been calculated, as also had been its magnitude, by Maxwell. It is extremely small, even with the greatest available intensity of light, so that very small currents of air are able to mask it. Crookes made experiments in exhausted vessels. Here, it is true, he actually found a pressure upon surfaces exposed to radiation; but it was not the pressure of radiation, but a new phenomenon caused by the gas molecules which were still present in large numbers in the best vacuum that could be obtained in those days. Crookes thus made his 'radiometer,' or light mill, in which a particular demonstration of the motion of gas molecules was given, but the pressure of light still remained masked on account of its smallness.

Later on, success in proving the pressure of light was obtained by improved pumps, increased intensity of light, and above all, by taking into account the radiometer effect; the pressure was measured and shown to be in agreement with Maxwell's theories; it has proved to be of no small importance for further reasoning. In cosmic space, where everywhere great intensity of radiation and large surfaces exposed to it occur, the effect of radiation pressure may be considerable; it may even outweigh the effect of gravitation.

Crookes was also the first to continue Bunsen's discovery of

to Würzburg, and in 1888 to Strasbourg. From 1894 to 1905 he was president of the Reichsanstalt. Kohlrausch exercised a very great influence for good upon the progress of electrical measurement; he was also responsible for the exact standard of electric current (ampère or weber) generally used to-day, and depending upon a measurement of the deposition of silver, made by him with the highest accuracy in the year 1881.

elements by means of the spectroscope; in this way he discovered thallium, in 1861. Inasmuch as in the same way elements which only occurred in traces in substances found on the earth's crust were gradually brought to light, the possibility then appeared of ordering all the known elements according to their atomic weight in such a manner that series of elements were found all having similar properties. This 'natural system of the elements' then immediately allowed the gaps to be seen by inspection, in which elements still undiscovered belong, and the number of these gaps is already approaching zero. The study of fluorescence by means of cathode rays, which was begun by Crookes, has shown itself in its modern form, in which invisible and high frequency fluorescence is used, to be a particularly effective method for the discovery of the last and still missing elements. The spectrum of this fluorescence, produced by means of crystal gratings, shows not merely the presence of new elements by new lines, but also betrays, by the position of the lines, the atomic number of the new element in the periodic system, whereby we are able to calculate directly the approximate value of the atomic weight, and also to derive many indications of the chemical behaviour of the element in question.

JOSEF STEFAN (1835-1893)
LUDWIG BOLTZMANN (1844-1906)

THESE two investigators made such important developments in the views initiated by Kirchhoff concerning the light and heat radiation of hot bodies, that their work served as the starting point from which the great bulk of our knowledge in this field up to to-day has been acquired. Kirchhoff had, by means of his law, established a definite connection between the emission and absorption of radiation. This already furnished a possibility, that in further investigating the peculiarities of emission, one might become independent of the particular qualities of the hot bodies emitting the radiation, and so be able to investigate the influence of temperature, for itself alone. The point was to determine this influence in the case of a completely 'black' body, that is to say, one which completely absorbed radiation of all wavelengths. Any body which does this has the maximum possible power of absorption, and in this respect all black bodies are alike, no matter of what they consist; for none can absorb more than everything. But in this case, Kirchhoff's law tells us that all black bodies must be alike as regards radiation also, and must also be possessed of the maximum possible radiating power, which will now no longer depend upon anything but the temperature. The emission of a completely black body must be investigated as regards its dependence upon the temperature, in order to find out the essential features of this dependence.

If we are to measure the whole radiation emitted by a body, the infra-red always present must not be overlooked; we cannot therefore make use of the eye; it is necessary to allow the radiation proceeding from the body to fall upon a black and sufficiently sensitive thermometer, such for example as a blackened thermopile, and to measure the corresponding rise in temperature; this gives us the correct measure of the total radiation, since this is completely absorbed by the black surface of the measuring instrument, the whole of its energy being turned into heat, and thus into a measurable form.

Measurements of this kind had already been published in Stefan's time (1879) from various sides. Stefan collected them all and made a critical comparison of them, paying particular attention to a possible effect of the heat conductivity of the air, which had in part been overlooked. Good measurements, for example by Tyndall, existed, in which an electrically heated platinum wire had been measured as regards its radiation from temperatures below redness up to the melting-point of the platinum. Stefan found from all such measurements without exception, that the total radiation from a black body is proportional to the fourth power of its absolute temperature (that is, measured from − 273°C); this rule has since been known as 'Stefan's law.' Stefan was also able to state, on the basis of some good measurements, the number of calories which one square centimetre of surface of a black body radiates in one second, and since the radiation arriving on the earth from the sun had likewise been measured in calories, he was able by means of his law to make for the first time a well-founded statement concerning the temperature of the sun. He found it to be approximately 6000° C (absolute), the sun being assumed to have the properties above defined of a black body, which agrees with all recent experience.

Boltzmann succeeded in making use of an imaginary experiment to bring Stefan's law, which had been derived from experience, into quantitative connection with Clausius' second law of thermo-dynamics, and with the pressure of radiation derived from Maxwell's theory (1884). This was of great importance, since Stefan's law suddenly ceased to depend for support only on the special observations which he used, and took its place in the firmly constructed theoretical structure formed by the theories we have mentioned. This had the greater significance inasmuch as not long after, Hertz was able to bring a further experimental confirmation of Maxwell's theory.

In this way, twenty-four years after Kirchhoff's law of emission and absorption had been firmly founded, Stefan's law could be added as a second and equally well-established law concerning the ether wave radiation of hot bodies. From this point progress was now much more rapid, the more so as Boltzmann had already left behind two essential

JAMES CLERK MAXWELL

WILHELM HITTORF

JOSEF STEFAN

WILLIAM CROOKES

preliminary investigations of wide importance, which continued to influence research until quite recently. By admirable and very extensive studies concerning the motion of gas molecules, he was able, basing himself upon the previous work of Clausius and Maxwell, to bring the concept of entropy, so important in connection with the second law of thermo-dynamics, into a definite relationship with the value of the probability (as already defined by Huygens) of the physical state of the group of bodies considered, whereby a new foundation was given for the application of the concept of entropy to further deductions. Furthermore, Boltzmann in the course of these investigations, and in the same year (1877), had also introduced a peculiar mode of calculation for discontinuous or quantum-like distribution of energy among gas molecules, which later became of especial importance when somewhat differently applied.[1]

The most important questions concerning the radiation of hot bodies remaining open after Boltzmann's time related to the distribution of the total radiation given by Stefan's law among the various wave-lengths in the spectrum. The simplest observation of a body gradually heated to incandescence, such as a piece of carbon or iron, already shows us that from the beginning of a red heat (about 600°C) we have the addition of red radiation to the hitherto invisible infra-red, then yellow appears, until finally at a white heat, the blue and violet also appear in the spectrum. The range of rays sent out thus gradually moves in the direction of shorter and shorter wave-lengths as the temperature rises. The law of this displacement could again be deduced by means of an imaginary experiment from already wellfounded knowledge; it was now necessary to make use, in addition to Clausius' second law of thermo-dynamics, and Maxwell's pressure of light, also of Doppler's principle. The

[1] See Boltzmann's scientific papers, edited by F. Hasenöhrl, vol. 2, page 167.

Zs

law states that the most intense wave-length radiated by a completely black body is inversely proportional to its absolute temperature.

There now only remained the question of how the longer and shorter waves always present along with the most intense radiation are distributed over the spectrum; in other words, the question of the exact distribution of the energy of radiation of a black body at any temperature over the whole spectrum. The answer to this question could not be given by the then existing knowledge and theory. In particular, new observations were necessary; the whole distribution of energy in the spectrum had to be examined in detail by means of the thermopile or similar apparatus, before further conclusions could be drawn. Here it was a matter of finding as a radiating body a trustworthy completely black body, and not only such a substance as soot or the like.

But Kirchhoff had already shown how to make such a body; it was necessary to use a hollow space having a small aperture; the aperture would behave like the surface of a completely black body. For if a ray of light of any wavelength falls from outside upon the hole, it enters the cavity, and if the internal walls of this are anything but perfectly reflecting, which can be easily arranged, at the worst only a very small part of the energy received escapes again from the hole, and this part can be still further diminished by making the hole smaller or by enlarging the hollow space. This blackness of an aperture opening into a hollow space can be seen at any cellar or in the pupil of the eye. But if the aperture behaves like a completely black body as regards absorption, it must do the same, according to Kirchhoff's law, also for emission. If, therefore, the hollow space is heated to a measured temperature, the aperture will always radiate exactly like a perfectly black body. This radiation can then be decomposed by the spectroscope, and its energy distribution over the spectrum measured.

These results, which are in the first place stated in the form of curves – for every temperature a curve with the wave-length as abscissa and the energy as ordinate – can only be interpreted if we succeed in summarising these curves by means of an equation in the manner of Descartes, which equation would then allow us, conversely, to deduce all the results of observation with the exactness of the original data. The production of this equation, and its interpretation, only succeeded after much trial, in spite of Boltzmann's excellent preliminary research, for the phenomena which are concerned in the radiation of a solid body are very complicated.

But the result was something entirely new and unexpected. As a basis we have a fact given by Maxwell's theory, which had already been confirmed by Hertz, and by our knowledge of the structure of matter from molecules and atoms, the unordered motion of which constitutes the heat contained in the body; the fact namely, that the radiating black body is to be regarded as a mass of electric oscillators or wave generators. These are able to respond to, and absorb, every possible wave-length, and are likewise able to emit all possible wave-lengths, in accordance with Kirchhoff's law; for this is the significance of the 'blackness' of the body. In radiating, they draw the necessary energy from the heat content of the body. The question was: how in these circumstances do the atoms operate as radiators, so that the actual result found by observation for the distribution of energy is produced ?

The answer towards which Boltzmann's calculations concerning discontinuous or quantum-like energy distribution had already pointed the way, was, if we now add the interpretation derived from the most recent research, as follows. Every atom radiates with the wave-length or frequency which is characteristic of it, but it does not radiate quantities of energy of any amount, but only multiples of definite quantities (quanta), so that it does not radiate until it has taken up this definite amount of energy (its energy-quantum or light-quantum) from its surroundings, whereupon this whole amount of energy is radiated in a train of waves. The magnitude of this energy is also proportional to the frequency of the atoms; simply proportional, that is to say. This is the substance of the 'quantum theory'; it was soon possible to enlarge it by the hypothesis that atoms altogether only deal with energy in quanta, whereby the magnitude of the quanta is always proportional to a frequency (the reciprocal of a characteristic time) which plays a part in the energy transformation, the proportionality being given by a factor which is always the same as that derived from the radiation of a black body.

We are here justified in speaking of a quantum theory, and no longer of an hypothesis, since many processes have already become capable of observation, in which this quantum-like transformation of energy in the atoms occurs, and particularly because the proportionality factor above mentioned gives rise to an exactly ascertainable quantitative connection between the spectral distribution of energy in black body radiation, and two natural constants which have already been accurately measured, namely the elementary electric charge, and the velocity of light.[1]

Josef Stefan was born in a small village near Klagenfurt in Karnten as son of poor parents of Slav race, who kept a small shop in which they sold food, and were unable to read or write. After he had first been occupied in carrying sacks of flour, it was made possible for him to go to school in Klagenfurt, where he made surprising progress, so that he was then allowed to study at the university in Vienna. After four years of study, the almost uninterrupted succession of his

[1] Further details of this development will be found in any modern text book of Physics. The lay reader may be referred also to C. G. Darwin's *The New Conceptions of Matter*, London, 1931. *The Structure of the Atom*, by E. N. da C. Andrade, London, 1927, is a full account for the physicist.

scientific papers already commenced, and these gradually covered all branches of physics. This versatility and the extraordinary thoroughness with which he undertook everything, also made Stefan very successful in the scientific education of teachers at that period, when, after himself having taught for seven years in a school, he was called at the age of twenty-five to be professor of physics at the university of Vienna, where he then remained for thirty-three years, until his death.

His lectures were of unusual clarity and perfection. As regards his experimental labours, the difficult investigations of the heat conductivity of gases deserve particular mention, since they afforded a new means of testing and confirming the kinetic theory of gases, which had become important in so many directions; up till then no sufficiently clear experiments and exact measurements had been made by anyone regarding the heat conductivity of gases. His mode of life was extremely simple and quiet. He was a silent man, but always anxious to assist honest scientific endeavour, regarding this as a self-evident patriotic duty. Boltzmann and Hasenöhrl were his pupils. He hardly ever went to scientific congresses, just as little as did Bunsen in his later days. His nature became more light hearted after he married, at the age of fifty-six. But he died only two years later after a short illness.

Ludwig Boltzmann was born in Vienna, and there studied mathematics and physics with Stefan, whose assistant he became. At the age of twenty-five he received a call as assistant professor of physics to the university of Graz, and shortly afterwards he became professor of mathematics at the university of Vienna. After he had carried out interesting experimental work concerning the dielectric constant of gases, in which a certain conclusion from Maxwell's theory was tested, he was called in 1876 to Graz as the chief representative of physics, where he remained until 1889, and carried out the greater part of his most important work. He then followed an invitation to Munich (he had refused Kirchhoff's chair in Berlin[1]), and finally became Stefan's successor in Vienna, in the year 1894. After six years there he moved for two years to Leipzig, but returned finally to Vienna, where the professorship had been kept open for him. It appeared as if his birthplace was after all the best for him and his family. Nevertheless, six years later, at the age of sixty-two years, he put an end to his life while on a journey.

Among all the great men of science which we have considered he was thus the first who took so great a distaste to life upon earth. Bodily suffering and periodic depression could not have been the sole causes of this; many others had already suffered from these complaints. We are here confronted with something hidden in the very depths of human development. In any case, Boltzmann was the last eminent man of science in Germany who appeared in the circles of physical congresses in order to give open expression to his opinion, which had been formed in the spirit of the great investigators of past times. This expression was often very strong, but nevertheless always met with a certain sympathetic reception.

HEINRICH HERTZ
1857–1894

HERTZ's confirmation of Maxwell's theory, as expressed in the latter's equations, dealt with this theory at its central point, namely the question of the existence or non-existence

[1] The offers in Berlin had been very attractive, but he did not feel comfortable there, for he liked to be free, and he had received a hint from a very important feminine quarter that he did not know how to behave properly at table.

of 'electric waves,' with their properties as clearly predicted by the theory.

If these waves could be generated and proved to exist, and if they possessed the required properties, there would be no further elements of doubt in the theory, and it could then be used as a safe guide in dealing with further questions. Hertz approached this goal, which in spite of all previous work was in his time still very obscure or indeed invisible, gradually at first, but finally at surprising speed, and showed that it not only actually existed, but was easily attainable with full certainty.[1]

At the same time, therefore, he had opened up a new world of phenomena and placed it at our disposal; he thus became the discoverer of electric waves and 'rays of electric force' (1888). The results of all this with regard to our knowledge of nature will be indicated further on; what resulted in the way of applications is more than well enough known to-day from wireless and broadcasting.[2] We will describe the

---

[1] It is very noteworthy that in the fifteen years between Maxwell's publication of his work *On Electricity and Magnetism* and Hertz's discoveries, a great deal had been wiitten about 'Maxwell's theory' and in particular concerning the electro-magnetic theory of light, and these subjects had been lectured upon at universities, yet not even the beginning of a way to the goal had been made clear, for people simply played about with Maxwell's equations, but not with the ideas of Maxwell or Faraday; it was a mathematical game and not scientific research that was pursued, and the results were sterile. Hertz was the first who not only understood the equations, and knew how to deal with them mathematically when necessary, but also saw the structure of ideas upon which they were based by their originator, and understood how to move about in it. The equations are, so to speak, merely ground plans of this structure, and are far from being actual inhabitable apartments; the latter can only be produced by the architect, who knows how to grasp the ideas which have been put into the ground plan.

[2] The telephone was invented by a Frankfurt teacher of physics, Philipp Reis, in 1860; it was only perfected considerably later, though this might have happened more quickly considering the state of scientific knowledge at the time. To-day, when there are plenty of scientifically educated technicians available, almost everything which the progress of science offers us is at once practically applied, as we see in the 'art' of broadcasting.

way taken by Hertz, and leave out of consideration previous reconnaissances.

The meaning of Maxwell's equations showed that the electric waves in question must proceed from electric oscillations, which W. Thomson (Lord Kelvin) had already taught us how to calculate in every respect, and the production of which had already been proved, for example in the discharge of Leyden jars. In experiments with the discharge of jars, the periodicity of the oscillations had been measured, by observing the discharge spark in a rotating mirror, whereby it was clearly seen that at every discharge a number of sparks passed to and fro, the intervals between these being half the period of oscillation of the discharge; this period could thus be measured. These frequencies might amount to hundreds of thousands a second.

If the transverse waves imagined by Maxwell proceed from these oscillations, and are propagated with the speed of light into space, the length of the waves – measured from crest to crest – would be easily calculated in the manner rendered clear by Newton for all waves, from velocity of propagation and frequency; they would be measured in miles, and this was much too long, not only as compared with the space in which the waves could be tested for and investigated; but also because the energy available in the discharge generating the waves was not nearly sufficient to allow any perceptible effect to take place, in view of the diminution of intensity with distance, over a range of miles; thus it was not possible to generate even a single such wave with an intensity sufficient to allow it to be detected.

This shows that, in order to render such waves perceptible, it would be necessary to generate much more rapid oscillations than those hitherto studied; for the wave-length changes inversely as the frequency. These more rapid oscillations were to be expected, according to Lord Kelvin's calculation, in the case of the discharge of smaller capacities

and with smaller self-inductions in the discharge circuit.
The fact that oscillations actually occur in such circumstances
was proved by Hertz satisfactorily in experimenting with
electric discharges, and this was the beginning of the series of
investigations, which led to complete success within two
years.

In order to prove the presence of the invisible waves, and
indeed the actual existence of rapid oscillations in conductors
excited by spark discharge, Hertz made use of resonance.
This is the phenomenon of sympathetic oscillation, which
was quite rightly recognised by Galileo in connection with
sound waves, and it only takes place when the resonator is
tuned exactly to an equal period of oscillation. In the case
of electrical resonance, the two conductors which have been
tuned to equal electrical period of oscillation are suitably
placed alongside one another. If one of them, the 'oscillator' or 'transmitter,' is then caused to oscillate by a spark
discharge, sparks are seen on the other, the 'resonator,' as a
sign that it also is oscillating in sympathy, and the presence
of an oscillation is thus proved. The fact that the propagation from oscillator to resonator takes place by means of
waves cannot be proved in this way; the phenomenon thus
observed can indeed be quite simply regarded as a process of
mutual induction, the oscillator being called the primary
conductor, and the resonator the secondary conductor.

In order to prove the existence of waves, the two conductors must be sufficiently far apart, which gives rise to the
difficulty that the effect falls off rapidly as the distance increases. Hence Hertz first attempted to conduct the oscillations along wires. Since in the case of this kind of conduction the velocity of propagation is that of light, as had been
shown by Gauss and Weber upon the introduction of the
electric telegraph, waves should be demonstrable in the
wires of the same length as are required by Maxwell's theory
to exist in the free state around the oscillator. This proof

of waves in wires was successful, the waves being made stationary in exactly the same way as is possible with waves in ropes or water, or with sound waves, by reflecting them back upon themselves; this is very simply accomplished in the case of electric waves in wires, by insulating the ends of the wires. Hertz was then able to demonstrate the presence of nodes and loops of electric force in the wires by means of tiny sparks, and double the distance from one node to the next gave, as in the case of all stationary waves, the wave length.

He then also succeeded in producing stationary waves of this kind without wires in free space between the oscillator and a reflecting wall, and likewise in measuring their wave-length. The equality of the wave-length with that found in wires would indicate an equal rate of propagation in the two cases, that is to say the velocity of light predicted by Maxwell. This important demonstration was a matter of some difficulty for Hertz, since the space available to him was too small for an undisturbed development of the waves; this proof only succeeded somewhat later (1893) when the experiments were repeated in a large hall.[1] A metallically conducting wall proved to be suitable for reflecting the waves, which also agreed with Maxwell's theory. We then have a node of electric force on the wall itself; no spark is there exhibited by the resonator. At the loops, where the sparks appear, the transverse nature of the waves can be proved; the electric force is at right angles to the direction of propagation of the waves. It was then also possible to study satisfactorily the waves freely radiated from the oscillator by means of the resonator, and compare them with the effects predicted by Maxwell's equation, full agreement being found.

[1] A full account of these investigations by Hertz himself is to be found in his work *Electric Waves*, translated by D. E. Jones, with an introduction by Lord Kelvin, London, 1893 and 1900. See also his *Miscellaneous Papers*, translated by D. E. Jones and G. A. Schott, London, 1896.

Thus waves were found in the ether, exactly similar to light waves in every respect excepting length. While the waves of visible light, measured by Fraunhofer and Fresnel, were of lengths measured in ten-thousandths of a millimetre, Hertz had obtained invisible light waves of a length measured in feet, or when he made his oscillator still smaller, in inches; and he was able to investigate their electric and magnetic force in detail. No further doubt could exist that the waves of visible light are also such waves of electric and magnetic force, although these forces, in view of the smallness of the waves, cannot be detected by ordinary means. A particular proof of the fact that the waves of electric oscillators have all the properties to be expected in long light waves, was given by Hertz in his well-known concave mirror experiment. He here obtained rays of electric force analogous to the rays from a searchlight, and he could demonstrate their refraction by a prism and measure it, everything being again found to agree with Maxwell's equations. In this way, not only was the electro-magnetic nature of light waves and of all phenomena of light proved and rendered perfectly clear, but it was also shown that all known electric and magnetic phenomena, including those of light, are processes of the same type taking place in one and the same ether, and only differing in their spatial arrangement, but in every case following Maxwell's equations. Maxwell's equations were now the summary of everything known up to that time concerning the ether.

Since then, ether waves of every length have become known, and can be generated. The waves measured in kilometres or centimetres and proceeding from electrical oscillators, link up, in a long scale of wave-lengths, with the longest infra-red waves which proceed from hot bodies, and so with the waves of the visible spectrum from red to violet; these are followed by the still shorter and again invisible ultra-violet, which are contained in the light of electric sparks and of the electric arc; finally we have the still shorter waves of the X-rays and gamma rays, the latter being given off by radio-active bodies, and their length being measured in millionths of a millimetre.

Since the correctness of Maxwell's equations was now placed beyond doubt, it was also clear that they should state correctly the limits of validity of single laws discovered earlier, such for example as Coulomb's law for the force between electrified bodies.   This law only relates to stationary electricity, as can be easily recognised.   If, for example, a quantity of electricity is situated not far from an oscillator, it is exposed to forces which can be more or less correctly calculated by Coulomb's law, in spite of the motion of the electricity upon the oscillator; but if the quantity of electricity is half a wave-length away from the oscillator, the forces have changed their direction; they would now be calculated by Coulomb's law wrongly, not only in respect of their magnitude, but also their direction.   This also renders clear the fact that 'actions at a distance,' determined in a fixed manner by the distance, are non-existent in these cases.   On the contrary, the forces exerted by the electricity, and proceeding from it, spread out into space, and require time for this purpose; and although the propagation takes place with the velocity of light when the motion of the electricity is very rapid, and particularly when it is of a to-and-fro character, as in the case of the oscillations, the delay becomes easily perceptible.   This propagation with delay is alone that which causes waves to be formed when electricity oscillates.   At the same time the forces, though Coulomb's law fails to represent them, are quite correctly given by Faraday's lines of force, only these lines of force must have ascribed to them the property of moving in a direction at right angles to their own direction, with the velocity of light.   The exact manner in which this happens, and the forms thereby assumed by the lines of force, are correctly given by Maxwell's equations, a fact of which Hertz brought as complete a proof as possible in the case of the oscillator.

We see that Maxwell's equations exhibit the behaviour of the electric and magnetic lines of force, their form and their motions, even in very complicated circumstances. These lines of force, which, when they are not closed in upon themselves (as are the magnetic lines always, the electric lines in waves), are linked at one end to positive, at the other to negative electricity, afford accurate and exhaustive pictures of states of space – of the ether – which are inseparably connected once and for all with the two kinds of electricity. For this reason Maxwell's equations appeared actually as the equations exhibiting the behaviour of ether, as far as that was known at the time.

This necessarily suggested that the behaviour of ether could be compared by means of these equations with the behaviour of matter. We know that in liquids and gases propagation of waves also occurs – these are known as sound waves – and that liquids and gases are also able to transmit pressure forces which likewise require time for their propagation. In order to compare the ether with a liquid, for example, it was a matter of comparing Maxwell's equations with the equations of hydro-dynamics, and seeing whether complete agreement exists when, for example, the electric forces are interpreted as displacements in the ether, or whether other interpretations also lead to agreement.

Mathematical investigations made for this purpose have all led to the result that striking similarity exists between ether and liquids or gases (or also solid bodies) in certain respects, but that complete agreement is wanting.[1] We can

[1] Particularly the profound investigations of Carl Anton Bjerknes, which were begun before the appearance of Maxwell's work, and were also not connected with it, 'On Hydrodynamic Forces at a Distance' were an important contribution to a clearer understanding of the mechanics of the ether. They appeared in English, in lectures given in 1905 at Columbia University by his son V. F. K. Bjerknes, and published under the title *Fields of Force*, London and New York, 1906.

certainly say that electric lines of force have striking simi-larity with vortex filaments or threads, and magnetic forces with lines of flow in liquids or gases – which gives a useful basis for considering many problems; but no material body can be described which would behave exactly as the ether of electro-magnetic fields of force, and of light. It has thus become clear that ether and matter are things very different from one another. The first does not consist of the atoms known since Dalton's time; the laws of mechanics, which are all derived from the observation of matter, cannot be applied to the ether at all, or at least only to a limited degree. The ether is not a mechanism of a material nature, but has peculiar properties which go beyond those given by our experience with matter, and can only be studied in con-nection with the ether itself.[1]

After Hertz's time, new information concerning the electro-magnetic field arrived when the investigation of cathode rays, begun by Hittorf and Crookes, led us to the knowledge that electro-magnetic fields also exist in the interior of the atoms of matter; and the question arises whether these fields also behave in accordance with Maxwell's equations. This question must be answered in the negative, since we have obtained knowledge of the quantum-like behaviour of the atoms, which goes back to Boltzmann. The ether inside the atoms does not however appear to behave entirely differently from that outside

[1] In this connection, Tyndall's description of his impression of Fara-day's remarks concerning the behaviour of ether, as regards lines of force, is very remarkable. Tyndall says that Faraday often used peculiar, and sometimes obscure, turns of speech, to describe this behaviour, as though he were not in a position to make use of expressions which would be intel-ligible to anyone with a knowledge of mechanics (Tyndall, *Faraday as a Discoverer*, London, 1807, page 88). Tyndall would surely be able to say to-day that Faraday's remarks were the more correct, since the mechanical mode of expression, which he did not use, could not be suitable for ex-pressing reality. It was simply the materialistic preconception, which is still not yet completely eradicated, which prevented Faraday's point of view, in spite of its correctness, from being followed.

the atoms,[1] but it shows certain peculiarities there, which are not contained in Maxwell's equations, and still form the subject of continued investigations.[2]

Thus Maxwell's equations also have their limits of validity; they relate only to electro-magnetic fields at a sufficient distance from atoms and to all phenomena thus taking place. It is remarkable in this connection that these fields – which alone were known before the discovery of cathode rays, and have been studied since Coulomb's time – never come into existence in any other way than by springing from the interior of atoms. No free electric charges are known which do not arise from atoms, within which the elementary particles of positive and negative electricity are found in the closest connection; if charges are to be collected, it is necessary somehow to separate sufficiently far from one another the two kinds of electricity, which are found nowhere else than in atoms, and by virtue of the lines of force which connect them, always strive to return there. When this is done, the field given by Maxwell's equations comes into being between the two kinds of electricity which have been collected in the form of a very large number of elementary charges.

The same is true of all magnetic forces which we use, even the strongest, such as those which drive electric railways. They can only be obtained by means of moving electricity obtained from atoms, or directly from atoms (best iron atoms) in which moving electricity is already present, all that is necessary being to arrange the fields of the many such atoms

---

[1] Accordingly, Maxwell's equations can be applied in certain respects to electro-magnetic fields within matter, for which purpose the constants (di-electric, magnetic, conductivity), appearing in the equations, are required. The dependence of the velocity of propagation (refractive index) upon the wave-length can be correctly calculated by the equations when the atoms or molecules of matter are regarded as electric resonators with their own period of oscillation (electro-magnetic theory of dispersion).

[2] A possible conception of the electric and magnetic lines of force which could also be applied inside the atom, will be found in my publication *Über Äther und Materie*, Heidelberg, 1911.

or molecules in such a way that they support one another in their external action, and do not become imperceptible externally by reason of want of order in their arrangement. In the same way, all kinds of ether waves, however different they may be, according to their length, in their peculiarities, effects, and mode of origin, nevertheless only come from matter. Matter is therefore without doubt the source of all electro-magnetic fields, and nevertheless these fields have other qualities at their place of origin in the interior of the atoms, than those revealed after they have escaped outside them. They show inside the atoms peculiar properties, which go beyond Maxwell's equations.

As regards the limits of validity, which are thus not wanting also in the case of Maxwell's equations, the course of our progress in knowledge is always such that laws are found and confirmed by experiment before the limits of their validity can be recognised. The latter, that is to say the conditions which must be fulfilled for complete validity, are often discovered much later, as our experience grows. Up to the present, the only principle possessing unlimited validity appears to be the energy principle of Robert Mayer.

In his experiments on electric oscillations and waves, Hertz had continually to observe small electric sparks produced by the resonators. In this connection he noticed that the sparks were longer when the light of another simultaneous spark, or any kind of ultra-violet light, fell upon the spark gap. He also discovered that it was chiefly or exclusively the negative electrode upon which the effect depended (1887). In this way he made a discovery, the further prosecution of which has likewise opened up a large field of knowledge: that of photo-electricity. As is always the case, the main matter was to produce the new phenomenon in the simplest possible form, which was very soon done (1888),[1] and then to follow it up with experiments of the

---

[1] Hallwach's effect.

Ludwig Boltzmann

Friedrich Hasenöhrl.

Heinrich Hertz

clearest possible type, which happened eleven years later. It then appeared that when air is excluded, so that the ultra-violet light acts only upon the metallic plate upon which it falls, the light causes free electrons to leave the metals. This is the fundamental process, the thorough study of which then not only brought understanding of the various forms of photo-electric effects, also in air and other gases, but led also to a very great deal of further knowledge.

It should first be mentioned that the phenomenon of 'phosphorescence,' which had been known for over a century, now could be understood, after G. Stokes had already recognised (1853) that these phenomena, as well as that of 'fluorescence,' consist in a transformation of light taking place inside the body concerned, and resulting in a change in the refrangibility, that is the colour, of the light. It now became evident, when phosphorescent material suited to proper experiments had been made, that the exciting light liberates electrons from the metallic atoms of the phosphorescent substance, and that it is the return of these electrons which causes the after-glow with change of colour, which is peculiar to the metal contained in the phosphorescent substance. This rendered it probable that the production and emission of light – for example also in metallic atoms in the Bunsen flame – is connected quite generally with displacements of electrons.[1] This fact, particularly in conjunction with our knowledge of the structure of atoms and the quantum-like manner of their operation, very considerably advanced our understanding of all processes of light emission and its excitation.

Furthermore, the photo-electric effect first enabled the

[1] Almost at the same time, studies of the 'canal rays' brought the same idea very close. I have given some historical data on an earlier occasion (*Quantitatives über Kathodenstrahlen*, new edition, 1925, pp. iv and v). The English reader should consult J. J. Thomson's *Conduction of Electricity through Gases*, Cambridge, 3rd Ed., 1928; Zworykin and Wilson's *Photocells*, London, 1930.

AAS

production of cathode rays of very low velocity – slow electrons – for pure experiments in a vacuum. By subjecting them to suitable electric forces, they could be accelerated or retarded at will, as was already known in respect of the rapidly moving electrons in discharge tubes, but they could also be brought to a complete standstill and caused to reverse their direction, so that no continuous emission took place. All this has become of the greatest importance in applications which opened up entirely new possibilities (complete rectification of alternating current, production of undamped electric oscillations, and almost unlimited amplification of weak alternating currents) particularly since success was obtained in producing these slow electrons in a complete vacuum in a still simpler manner, and at the same time in unlimited quantity, by means of their emission from hot wires. It had long been known that incandescent bodies allow electricity to escape; however, as long as the observations were only possible in air, no results capable of leading to an explanation could be obtained.

Here also it was first of all necessary to arrange the experiments clear of all the complication produced by the presence of gas. The fact that these experiments were only possible at so late-a-date, or at least could only be carried out with certainty, was due to the continual emission of large quantities of gas from all incandescent bodies, which results in the destruction of the vacuum; only tungsten – the suitability of which was not discovered until 1913 – does not exhibit this effect, and hence allows of perfectly definite experiments being made and of satisfactory practical application, in connection with which the experience already gained with photo-electric effects was available.

The discovery of photo-electricity by Hertz came as a surprise; electrified bodies and the possibility of exposing them to ultra-violet light, had already been in existence for three quarters of a century, but no one had appeared to show us that something was hidden behind it all. Even Hertz had only arrived at the discovery by way of his confirmation of an existing theory, which had been derived from Faraday's great store, the man who always searched with confidence in the unknown.

Heinrich Hertz was the son of a lawyer and senator of Hamburg, and partly of Jewish blood. He attended the school in his birthplace, and then first studied technical science, but soon decided to devote himself entirely to physics.[1] After three years of study in Munich and Berlin, he became assistant to Helmholtz in Berlin, entered Kiel University three years later, and only two years afterwards, having in the meantime published a number of papers on different parts of physics, he received a chair at the technical high school in Karlsruhe. He there carried out his investigations on electric waves, and he also soon married. In the year 1889 he went to Bonn as the successor of Clausius. While there he published his remarkable work on the *Principles of Mechanics, presented in a new form*.[2] From 1892 onwards he became afflicted with an incurable disease, which led to his death from blood poisoning at the age of only thirty-seven years.

FRIEDRICH HASENÖHRL

1874–1915

FROM the time of Faraday's discovery of self-induction, this phenomenon could already be regarded as an inertia effect. For the electric current in a wire behaves, on account of self induction, like a current of water in a pipe. When it is set

[1] See *Heinrich Hertz, Erinnerungen, Briefe, Tagebücher*, collected by his daughter Dr. Johanna Hertz, Leipzig.
[2] Trans. by D. E. Jones and J. T. Walley, London, 1899.

in motion a delay occurs – in the case of the electric current as a consequence of the opposing electromotive force, in the case of the water on account of mechanical inertia; and when the current is interrupted we perceive a striving for it to flow on – again on account of the electromotive force of induction in the wire and on account of inertia in the water.

This phenomenon of inertia in the case of electric current must not be interpreted, at any rate not for the most part, as inertia of electricity in the wire. For the magnitude of the inertia – the self-induction – is very different in the same wire according to the manner of its winding, and to its surroundings; it is for example much larger in a coil of wire, than when the same wire is stretched out straight – and it becomes very much greater still if iron is introduced into the coil. If self-induction is to be regarded as a phenomenon of inertia, it would be necessary to look for the inertial mass, not in the current carrying conductor, but in the lines of force belonging to the current; for it is these that determine the self-induction.

This view remained unlikely, as long as Faraday's idea of lines of force still appeared as of minor importance; but it suddenly became very obvious, when Hertz's result had shown these lines of force to be without any doubt representative of states of space or ether, which make up the most essential feature of all electrical and magnetic phenomena. From that time on, an increasing number of attempts were made to connect inertia or mass with self-induction, that is with the states or processes in the ether represented by Faraday's lines of force; the idea being that perhaps all inertia – all mass – even that of ordinary material bodies, might be shown to be simply the mass of the accompanying ether.[1]

---

[1] In an analogous way Hertz developed his mechanics (1894) by taking account of mass in the ether, though in another way; the object was not to refer tangible mass, but all forces of nature, to hidden ether masses.

These endeavours only found a firm basis when the attempt was made to connect up with mass or inertia, not the ether, which is difficult of access to us, but energy, defined by Robert Mayer's well-founded and repeatedly confirmed law of conservation, and always susceptible of accurate measurement, in spite of the great variety of forms which it assumes. Maxwell's theory already suggested that attention should be directed, not to the ether of the lines of force, but to the energy connected with them. For electromagnetic energy can be calculated according to Maxwell's equations from the course of the existing magnetic lines of force, whereby each element of space traversed by lines of force receives a definite fraction of the energy, so that the energy actually appears as distributed in the ether of the lines of force, almost as if this ether, and the energy, were perhaps one and the same thing. Hence it might reasonably be attempted to ascribe to the energy of the field of force of an electric current, or of moving electricity generally, the mass or inertia which appears in the phenomenon of self-induction. In this way the idea arose that energy – at least electro-magnetic energy – has mass, that it possesses inertia, that it resists change of velocity.

It was first attempted to carry out this idea in the simple case of a moving electron, that is to say, pure electricity without matter, which could be done by means of Maxwell's equations; but the attainment of a satisfactory result was prevented by the necessity for making certain assumptions about the electron which went beyond existing experience. Hasenöhrl was the first who put aside the electron, about which too little was known, and turned to those electromagnetic fields which can exist without either matter or electricity. These are the fields of force of the electric waves realised by Hertz, about which it was also known that they behaved according to Maxwell's equations. The fact that these waves carry energy, as do all ether waves of whatever wave-length, including those of light, is beyond all doubt; the energy can always be measured in the form of heat by causing the radiation to be absorbed in a suitable body. It was a matter of calculating the mass of this energy in the case of these waves.

Hasenöhrl carried out this calculation by means of an imaginary experiment with a hollow space, which was pictured as filled with ether waves and given an acceleration. Considerations, in which Maxwell's pressure of light played an important part, this pressure being exerted by the radiation enclosed in the hollow space upon the walls of the same, allowed that part of the inertia due to the enclosed radiation, that is to say, the mass of the energy of the radiation, to be calculated. This mass is shown to be equal in grams to the quantity of energy (measured in Gauss's absolute unit) divided by the square of the velocity of light.[1] This is Hasenöhrl's important and highly remarkable result.

In this way the mass of electro-magnetic energy could for the first time be calculated in a well-founded manner. But the assumption could immediately be made that this calculation ought to hold for every form of energy; for otherwise mass would disappear or appear out of nothing in the course

[1] Hasenöhrl's imaginary experiment and calculation of the same (*Berichte der Wiener Akademie*, Vol. 113, 1904, and *Annalen der Physik*, Vol. 15, 1904 and Vol. 16, 1905) is more or less complicated. I have set some value on the most completely attainable simplicity (*Äther und Uräther*, Leipzig, 1922, pp. 41, 42), and I find that the same was done before me but after Hasenöhrl, on another side; but Maxwell's pressure of radiation always remains the chief point. A simplification of the way by which an already known result is reached is naturally a comparatively easy matter; but it increases, and even opens up for the first time, our insight into essential features. We find in Hasenöhrl a factor 4/3 multiplying the mass of energy, which factor we left out above, as being close to unity and in accordance with the result of the simplified view; factors of this kind, which on account of the novelty of the matter, cannot be calculated at first with the completest possible approximation to reality, also appear in allied cases, such as the kinetic theory of gases. Some uncertainty in regard to them makes no difference to the main result, and they can then be corrected later, often with the recognition of further new truth.

of the continual transformation of energy, and this would
contradict our experience in mechanics. The masses of
energy entering into ordinary processes are however very
small; for the square of the velocity of light which enters
into their calculation in the denominator is a very large
quantity. Nevertheless, the sun, for example, loses energy
in the form of its radiation to the extent of five million tons
of mass per second, of which about two kilograms fall on the
earth, and these suffice to accomplish all known effects of
energy (with the exception of volcanic energy and that of the
tide).

If we assume Hasenöhrl's result to hold for all forms of
energy, it is the more remarkable, since energy now appears
exactly like a substance; it is not only invariable in its quantity, as Robert Mayer had already shown, but it also possesses mass. With or without this generalisation to cover
all forms of energy, Hasenöhrl's result was as well founded
as any new form of natural knowledge can be; it needs continual further proof in its application, until it is rendered
sufficiently secure, and the limits of its validity are known.
The applications of this idea have already progressed very
far to-day, although almost entirely in the names of other
people – but nowhere has any contradiction with experience
been found, but on the contrary continual confirmation.
Indeed, Hasenöhrl's result could be further extended to
show that, associated with the mass of energy, there is a corresponding *weight*, so that energy is subject to gravitation
like all masses, and further even, that the masses of tangible
and material bodies are to be regarded only as energy masses,
and the weights of these bodies as the weights of the energy
they contain.

In this connection we may make the following further
remarks.

A confirmation of the result was obtained by swift cathode
rays. If we regard kinetic energy as likewise possessing

mass like that of the electro-magnetic energy of light waves, the total mass of a moving body must increase when its velocity is increased, since there is added to the mass which it possesses in a state of rest the ever increasing mass of the kinetic energy which it contains. If this increase of mass is calculated, by means of Hasenöhrl's result, with the addition of Newton's second law of motion,[1] we find that the increase of mass only becomes perceptible at very high velocities, approaching that of light; at the velocity of light it would actually become infinite. In order to be able to test whether the calculated increase is actually present in reality, it was necessary to find masses which could be given very great velocities, which velocities could also be increased at will.

Such masses exist in the cathode rays, which are quickly moving electrons. In the course of measurements of great difficulty and increasing refinement, velocities were investigated which reached to within a few per cent. of that of light, and the increase of mass was found to be exactly that calculated from the kinetic energy.[2] This is also the particular reason derived from experience, for assuming that the calculation of the mass of energy is true for all forms of the same. This shows that the velocity of light cannot be exceeded by an electron, and therefore no doubt also not by matter, which is built up of elementary electric charges; for in order to attain the full velocity of light, an infinitely great expenditure of energy would be necessary.

The proportionality between mass and weight already

---

[1] I carried out the very simple calculation (*Äther und Uräther*, Leipzig, 1922, pp. 48, 49) in order to destroy if possible the impression which is widely prevalent that a peculiar, very complicated, and not easily understood theory, originated by Hasenöhrl, is necessary to it. I later found that Hasenöhrl himself already gave this simple calculation (Stark's *Jahrbuch*, vol. 6, 1909, p. 501), in which he refers to another author, who carried it out in 1908 after his fundamental publication. It is still more remarkable that even to-day, when these results are used, no mention is made of Hasenöhrl.

[2] Here we must replace Hasenöhrl's factor $4/3$ by 1, as we can also see from the simplified calculation from the pressure of light.

recognised by Newton at once suggests the question whether the mass of energy possesses a corresponding weight in this proportion, whether, therefore, energy is subject to gravitation. This question has been answered in two ways in the affirmative. In the first place, a deviation of the light from fixed stars when passing close to the sun has been observed at eclipses. The rays of light are bent towards the sun, and as far as the difficult observations, which are influenced by refraction in the sun's atmosphere, can show, about to the extent that would be expected in the case of any body subjected to gravitation and projected with the velocity of light close past the sun. The energy mass of light thus appears to exhibit, as far as we can see, a gravitation equal to that of the mass of all heavy bodies.

On the other hand, radio-active bodies, such as uranium, were used as the bobs of pendulums, in order that in their case the proportionality between mass and weight could be tested as in the pendulum experiments of Galileo and Newton, and this proportionality was found to be confirmed. In this case a distinctly sensible part of the mass of radio-active atoms consists without doubt of energy; for these atoms continually emit great quantities of energy in the form of the radiations and heat which they develop. If this part of the mass of the atoms did not possess a corresponding weight, the pendulum experiments would have shown a discrepancy, which however was not the case.

According to these results, which were obtained in connection with Hasenöhrl's investigations, energy behaves exactly like a substance. It does not alter its quantity, as Robert Mayer showed; it possesses inertia and also a corresponding weight. This immediately suggests that its distribution in space should be investigated in each individual case, and this can at once be carried out. We find that energy in all its forms always exists only in electro-magnetic fields of force. This allows us to assume that the whole mass of all substances, of all atoms including the non-radio-active, is only energy mass, so that also all ordinary weight is only energy weight. For our experience with cathode rays has shown strong electro-magnetic fields of force to exist in the interior of all atoms, and even the centres of these fields of force, the negative and positive elementary charges, are themselves to be regarded as concentrations of energy. It was further hardly to be doubted that the almost exact whole numbers of the atomic weights is the result of the atoms containing whole numbers of elementary quantum pairs,[1] while the – generally small – departures from whole numbers depend upon differences in the content of energy. and hence the weight, of the electro-magnetic fields of the pairs.[2]

If we accordingly regard ether and matter from this point of view, the distinction between the two, which together make up the material world, may appear to be obscured; for in both of them energy and hence mass and weight have been found, although in very different degrees. We have taken away from matter its hitherto imagined distinction from ether. Matter appears according to the facts we have just summarised only as a special case of energy, and we now have to set side by side not matter and ether, but rather energy and ether.

Of matter we must say that its atoms represent enormous concentrations of energy, which however for the most part must be looked upon as untransformable. Only the heaviest and radio-active atoms emit energy of themselves,

---

[1] Such a pair, consisting of a positive and negative elementary charge with the corresponding field of force, regarded as a fundamental constituent of all atoms, has also been called a 'dynamid.' Hence we may, at the present time, regard it as established that gravitation and inertia are properties belonging to energy and to it alone. These properties were at first discovered only in the case of matter, because its atoms are enormous accumulations of energy.

[2] A full account of these matters will be found in my paper *Über Energie und Gravitation*, Heidelberg Academy 1929.

and disintegrate in the process; otherwise all atoms are able to take up energy from their surroundings, and to give it up again, and as we have found, they do this by quanta. We have likewise already remarked that all energy existing outside the atoms (electro-magnetic waves and all electromagnetic fields of force) is only derived from atoms. As distinctive of matter there remains the positive elementary quantity of electricity, which appears to be a fundamental constituent of all atoms, inasmuch as none of these positive quantities, as distinguished from the negative charges, the electrons, can be removed from an atom without destroying the same, and transforming it into another kind of atom. The positive quantity also carries a very much larger mass – that is to say energy – than the negative, a mass over one thousand times as great, if the lightest atom, that of hydrogen, consists of only one positive and one negative elementary charge, as would appear from our present knowledge.

If we cast our eyes over the discoveries made concerning ether from Huygens to Hasenöhrl, we still notice that our knowledge is very incomplete. Originally, in Huygens' case, the ether appeared as the medium in which the waves of light are propagated, the knowledge of which was very greatly extended by Young, Fraunhofer and Fresnel. Then, thanks to Faraday, Maxwell, and Hertz it was generally recognised as the medium of all electro-magnetic forces, which are also the essential feature of light-waves. Later on, the interference experiment continually refined since the time of Fresnel,[1] together with observations of double stars, showed that the ether cannot be assumed to be either stationary or in motion in the whole of cosmic space, but

[1] We refer particularly to the Michelson experiments; but other electromagnetic experiments belong in the same category, and these have often been summarised, and will be found specially treated in my *Äther und Uräther*. A further interference experiment, which has hitherto been far too little followed up, appears to show that the ether does not take part in rotary motions of matter. The beginning of all these experiments goes back to Maxwell (see *Nature*, 29th January, 1880, page 314 ff.).

that it – as the medium in which light and all electro-magnetic fields are propagated with the velocity of light–shares to a corresponding degree the motion of every heavenly body, such as the earth, and indeed of every atom of matter to which it is in close proximity. Finally, after the steps forward initiated by Hasenöhrl, new questions re-garding the ether can be asked: does it not appear necessary – if every atom has its own ether – to ascribe to every quantity of energy its own quantity of ether, as being always attached to it, quite independently of whether it is energy within or without matter ? And does not the ether then appear altogether simply as an appurtenance of energy, which accompanies every quantity of energy in a large radius around it, decreasing in power and reaching as far as the gravitational effect of the energy reaches, that is to say – as far as we know – to infinity ? But would this not then immediately make it clear, as we have known to our aston-ishment since Newton's time, that every atom is at all times everywhere present with its gravitational effect ? Every ray of light would then – in accordance with its energy con-tent – carry with it its own ether, moving with the velocity of light through the surrounding ether. The ether would then by no means be a uniform medium filling all space uni-formly; but all space would nevertheless be filled with ether. The *Uräther* existing in cosmic space far from all heavenly bodies and there determining the velocity of light, would be the totality of all the various parts of ether, more or less equally represented, belonging to the energies (masses) present all around.

Space and ether appear, in view of such considerations, still less identical than ever. The ether is not space; it may be described as the secret of space, especially when we take into account the phenomena of life, caused by mind, but also taking place in space, together with the fact that every-thing in nature is always found to act as a single unity.

We see that the connection between space, ether and energy (to which matter also belongs), and gravitation, which at present can only be stated in the form of suppositions or questions, shows us to what extent we are surrounded by the vast unknown, which has only been reduced but little since the time of Pythagoras. But the fact that questions of this kind can, thanks to Hasenöhrl, be stated in so definite a manner, as to lead to new ideas, means that we stand at the beginning of further progress, which will occupy the great men of science of the future, and will without doubt lead to further surprises, inasmuch as nature will again speak to us, and these surprises will perhaps be very unlike what we now suppose, for this is what has happened in the course of the centuries, through the work of the great men whom we have considered.

Friedrich Hasenöhrl was born, like Boltzmann, in Vienna. His father was a lawyer, his mother came of an old military family. He first entered an aristocratic school, later the *Gymnasium*, and then studied natural science and mathematics at the university of his birthplace, particularly as a pupil of Stefan and Boltzmann. Before the conclusion of his university course he had already finished several mathematical investigations, which earned him the warm approbation of his teachers, and later also some experimental work. He soon married, and became a *privatdozent* in Vienna. After six years he became professor at the Viennese Technical High School, and a little later Boltzmann's successor at the university. He was only able to work there for eight years, under favourable circumstances.

Then the war broke out, and Hasenöhrl immediately volunteered. He was everywhere at the front, first at the defence of Przemysl, then in the Tyrolese mountains, which he knew and greatly loved. After a bullet wound, which was more or less healed, he again went to the front, and fell in the second year of the war at Vielgereuth, only forty-one years old.   He was of a simple, kindly and modest nature. Only when it was a matter of self-sacrifice was he to be found in the front ranks; when it was a matter of gaining rewards, he was in the background, even when he was the person most concerned, as we see in the most astonishing manner in his writings.   He loved music and his violin, as Galileo his lute; he was very fond of his family and as extremely modest and thankful for the smallest service as Kepler; the Alps were his beloved place of recreation as with Tyndall; in his work we find the thoroughness of Stefan and Boltzmann.

NAME INDEX

SUBJECT INDEX